WORLD HEALTH ORGANIZATION
INTERNATIONAL AGENCY FOR RESEARCH ON CANCER

 GREEN COLLEGE, OXFORD

# INTERPRETATION OF NEGATIVE EPIDEMIOLOGICAL EVIDENCE FOR CARCINOGENICITY

Proceedings of a Symposium held in Oxford, 4–6 July 1983

EDITORS

N.J. WALD & R. DOLL

IARC Scientific Publications No. 65

INTERNATIONAL AGENCY FOR RESEARCH ON CANCER
LYON
1985

The International Agency for Research on Cancer (IARC) was established in 1965 by the World Health Assembly as an independently financed organization within the framework of the World Health Organization. The headquarters of the Agency are at Lyon, France.

The Agency conducts a programme of research concentrating particularly on the epidemiology of cancer and the study of potential carcinogens in the human environment. Its field studies are supplemented by biological and chemical research carried out in the Agency's laboratories in Lyon and, through collaborative research agreements, in national research institutions in many countries. The Agency also conducts a programme for the education and training of personnel for cancer research.

The publications of the Agency are intended to contribute to the dissemination of authoritative information on different aspects of cancer research.

Distributed for IARC by
Oxford University Press, Walton Street, Oxford OX2 6DP

London  New York  Toronto
Delhi  Bombay  Calcutta  Madras  Karachi
Kuala Lumpur  Singapore  Hong Kong  Tokyo
Nairobi  Dar es Salaam  Cape Town
Melbourne  Auckland

and associated companies in
Beirut  Berlin  Ibadan  Mexico City  Nicosia

Oxford is a trade mark of Oxford University Press

Distributed in the United States
by Oxford University Press, New York

© International Agency for Research on Cancer 1985

ISBN 0 19 723065 2
ISBN 92 832 1165 0 (Publisher)

All rights reserved. No part of this publication may be reproduced, stored in a retrieval system, or transmitted, in any form or by any means, electronic, mechanical, photocopying, recording, or otherwise, without the prior permission of Oxford University Press

**PRINTED IN SWITZERLAND**

# CONTENTS

| | |
|---|---:|
| Organizing Committee | vii |
| Participants | vii |
| Editors' Note | ix |
| Foreword | 1 |
| **Introduction: Purpose of Symposium** | |
|   R. Doll | 3 |
| **Statistical considerations** | |
|   N. E. Day | 13 |
|   Conclusion | |
|     P. Armitage | 29 |
| **Oral contraceptives and breast cancer** | |
|   Laboratory evidence | |
|     P. Shubik | 33 |
|   Epidemiological evidence | |
|     M. P. Vessey | 37 |
|   Conclusion | |
|     B. MacMahon | 49 |
| **Hair dyes** | |
|   Laboratory evidence | |
|     W. G. Flamm | 53 |
|   Epidemiological evidence | |
|     L. Kinlen | 57 |
|   Conclusion | |
|     P. Fraser | 67 |
| **Hydrazine** | |
|   Laboratory evidence | |
|     J. R. P. Cabral | 71 |
|   Epidemiological evidence | |
|     N. J. Wald | 75 |
|   Conclusion | |
|     L. Kinlen | 81 |
| **Formaldehyde** | |
|   Laboratory evidence | |
|     W. G. Flamm & V. Frankos | 85 |
|   Epidemiological evidence | |
|     E. D. Acheson | 91 |
|   Conclusion | |
|     O. M. Jensen | 97 |

**DDT**
  Laboratory evidence
    J. R. P. Cabral .................................................................................. 101
  Epidemiological evidence
    J. Higginson ...................................................................................... 107
  Conclusion
    T. W. Anderson ................................................................................ 119

**Saccharin/Cyclamates**
  Laboratory evidence
    P. Shubik .......................................................................................... 125
  Epidemiological evidence
    B. K. Armstrong .............................................................................. 129
  Conclusion
    D. Krewski ....................................................................................... 145

**Phenobarbital**
  Laboratory evidence
    J. R. P. Cabral .................................................................................. 151
  Epidemiological evidence
    B. MacMahon ................................................................................... 153
  Conclusion
    R. Saracci ......................................................................................... 159

**Isoniazid**
  Laboratory evidence
    P. Shubik .......................................................................................... 163
  Epidemiological evidence
    T. W. Anderson ................................................................................ 165
  Conclusion
    J. Higginson ...................................................................................... 177

**Nitrates**
  Laboratory evidence
    W. G. Flamm ................................................................................... 181
  Epidemiological evidence
    P. Fraser ........................................................................................... 183
  Conclusion
    B. K. Armstrong .............................................................................. 195

**Beryllium**
  Laboratory evidence
    W. G. Flamm ................................................................................... 199
  Epidemiological evidence
    R. Saracci ......................................................................................... 203
  Conclusion .......................................................................................... 221

**General conclusions**
  R. Doll ................................................................................................. 225

Index of authors ..................................................................................... 229
Subject index .......................................................................................... 231

## ORGANIZING COMMITTEE

R. Doll
R. Saracci
P. Shubik
M. P. Vessey
N. J. Wald

## PARTICIPANTS

E. D. Acheson[1]
Director and Professor of Clinical Epidemiology, MRC Environmental Epidemiology Unit, Southampton General Hospital Southampton S09 4XY, UK

T. W. Anderson
Professor and Head of Health Care and Epidemiology, University of British Columbia, Vancouver, BC V6T 1W5, Canada

P. Armitage
Professor of Biomathematics, University of Oxford, UK

B. K. Armstrong
Director, NH and MRC Research Unit in Epidemiology and Preventive Medicine, The Queen Elizabeth II Medical Centre, Nedlands, Western Australia 6009, Australia

J. R. P. Cabral
Scientist, Unit of Mechanisms of Carcinogenesis, International Agency for Research on Cancer, Lyon, France

N. E. Day
Head, Unit of Biostatistics and Field Studies, International Agency for Research on Cancer, Lyon, France

R. Doll
Warden of Green College, Oxford and Honorary Director, Imperial Cancer Research Fund Cancer Epidemiology and Clinical Trials Unit, Radcliffe Infirmary, Oxford, UK

W. G. Flamm
Associate Director, Toxicological Sciences, Bureau of Foods, Food and Drug Administration, Washington DC, USA

P. Fraser
Senior Lecturer, Epidemiological Monitoring Unit, London School of Hygiene and Tropical Medicine, London, UK

---

[1] Present address: Chief Medical Officer, Department of Health & Social Security, Division IR1A, Alexander Fleming House, Elephant & Castle, London SE1 6BY, UK

J. Higginson
Director, Universities Associated for Research and Education in Pathology Inc, Bethesda, MD, USA

O.M. Jensen
Director, Cancer Registry, Institute for Cancer Epidemiology, Copenhagen, Denmark

L. Kinlen
Director, CRC Cancer Epidemiology Unit, Edinburgh, Scotland, UK

D. Krewski
Chief, Biostatistics and Computer Applications, Health and Welfare, Ottawa, Canada

B. MacMahon
Professor, School of Public Health, Harvard University, Boston, MA, USA

R. Saracci
Head, Unit of Analytical Epidemiology, International Agency for Research on Cancer, Lyon, France

P. Shubik
Senior Research Fellow, Green College, Oxford, UK

M.P. Vessey
Professor of Community Medicine, University of Oxford, UK

N.J. Wald[1]
Deputy Director, Imperial Cancer Research Fund Cancer Epidemiology and Clinical Trials Unit, Radcliffe Infirmary, Oxford, UK

---

[1] Present address: Professor, Department of Environmental and Preventive Medicine, St Bartholomew's Hospital Medical College, University of London, UK

# EDITORS' NOTE

The Editors were aware that since the time of the meeting, new studies had been reported on a number of the compounds considered, but since these could not be discussed by all the members of the Symposium, they have not been included in these Proceedings.

# FOREWORD

The role of the International Agency for Research on Cancer, as a research institution committed to public health, is to generate and disseminate information useful for the prevention of human cancer. Following a request received in 1968 to provide information on environmental chemical carcinogens, the Agency has devoted one of its most important programmes to identification of environmental carcinogens and to evaluation of the probability that exposure to them may lead to cancer in humans. Through this programme, which is centred on the production of a series of *IARC Monographs on the Evaluation of the Carcinogenic Risk of Chemicals to Humans,* over 700 chemicals, groups of chemicals or mixed exposures have been evaluated. More than 12 years after the first request to prepare a list of human carcinogens, the Agency extracted from its *Monographs* series a group of chemicals and occupational exposures that are evaluated as being causally associated with cancer in humans, and another group of chemicals and industrial processes that are evaluated as being probably carcinogenic to humans.

The Agency welcomes any input that could better the criteria for evaluating data on carcinogenicity, as this will help to upgrade its contribution to cancer prevention. The Agency therefore supported the initiative taken by Sir Richard Doll and Dr P. Shubik to call a symposium to discuss one of the most difficult and controversial issues in cancer epidemiology—the interpretation of negative epidemiological results. The Agency is glad to be able to publish these proceedings, which constitute an important contribution to the evaluation of epidemiological data. It should be stressed, however, that the resulting publication is completely separate from the *IARC Monographs* series. For instance, each volume of the *IARC Monographs* is the responsibility of a large working group of experts in many fields of research related to the evaluation of carcinogenic risk; in the present publication, each article is individually authored and is therefore the sole responsibility of the authors and the editors. The categorization of evidence proposed in the introduction to this volume as a working tool for the participants of the Symposium is in no way connected with the categories used in the *IARC Monographs.*

The chemicals chosen for consideration in this Symposium had all, except for nitrates, been evaluated within the *IARC Monographs* series, and all may be seen as 'problem compounds', in the sense that the available epidemiological data did not reveal a causal relationship between exposure and human cancer, nor did they provide clear evidence for the absence of such an effect. In particular, the evidence provided by data from long-term tests in experimental animals was sometimes at variance with the epidemiological evidence.

Our presently limited knowledge of the mechanisms of carcinogenesis does not allow a direct extrapolation from experimental data to the human situation, as this would be scientifically unjustifiable. For this reason, the Agency has adopted a prudent attitude (supported by the advice of many experts), which is to consider that in the absence of adequate human data, and for practical purposes, chemicals for which experimental results provided sufficient evidence of carcinogenicity (i.e., evidence of a causal association), should be regarded as if they presented a carcinogenic risk for humans.

By definition, and by the nature of the facts, data obtained from studies in animals, strong as they may be in demonstrating the absence of carcinogenic effect, will never invalidate a positive epidemiological study. By the same nature of facts, the reverse should be true—that is, an epidemiological study or series of studies showing that an exposure did not cause an increased risk of developing cancer should overwhelm any experimental evidence of carcinogenicity. In reality, the conditional in the preceding sentence indicates that it could do so, if and when the negative evidence provided by epidemiological studies is adequate and convincing, an ideal situation that is very rarely reached. The participants in the present Symposium have considered, on a case-by-case basis, what might constitute 'adequate and convincing' evidence.

What remains to be seen, therefore, is the extent to which epidemiological studies can be used to conclude that a compound does not represent a carcinogenic risk to humans, even if there is clear evidence of carcinogenicity in experimental studies. As Sir Richard remarks, 'No hard and fast criteria can be laid down that will automatically lead to an appropriate conclusion in all circumstances'. In the pages that follow, an attempt has been made to distinguish between the different degrees of evidence that may be provided by epidemiological data. The presentations in this volume may not provide elements that would justify a change in the attitude of the Agency, but they contribute to the clarification of several aspects of the evaluation of carcinogenicity data and point to the necessity of obtaining better information covering, in particular, individual exposure levels.

I should like to congratulate the initiators of this Symposium as well as all the participants for their contributions, which have undoubtedly helped to deepen the discussion on the validity and the limits of the data on which a qualitative, and ideally also a quantitative, assessment of human risks has to be based.

L. Tomatis, M.D.
Director,
IARC

# PURPOSE OF SYMPOSIUM

R. DOLL

*Imperial Cancer Research Fund
Cancer Epidemiology and Clinical Trials Unit,
Gibson Building, Radcliffe Infirmary, Oxford OX2 6HE, UK*

## PREFACE

The opening session was devoted to discussion of the purpose of the Symposium and the way in which subsequent sessions could be made most fruitful. A number of suggestions were made for improvement of the paper that had been prepared for discussion at this session, and these have been taken into account in the text that follows. The exact wording remains the responsibility of the author (RD), but the general tenor of the paper, as it now appears, was acceptable to the participants in the Symposium as a whole.

## INTRODUCTION

The progressive reduction in mortality rates that has taken place over the last 50 years and the growth of our ability to control the spread of infectious diseases and to treat them effectively have, in many countries, concentrated the attention of those concerned with public health on the main killing diseases of middle life and early old age and, in particular, on cancer and ischaemic heart disease. These, between them, are now frequently responsible for more than half the loss of all expectation of life under 85 years of age—the former being responsible, in England and Wales, for slightly more than the latter (Doll, 1983).

So far as cancer is concerned, it is now common ground, for all who have been engaged in its study, that the great majority of cases are in principle preventable, in the sense that it should be possible to reduce the risk of developing the disease at a given age by some 80 or 90% (or, rather, reduce by this amount the risk of developing the sum of the many disparate diseases that are grouped together under this generic head). There is, however, less unanimity about how this objective can be reached. With one strategy we seek to understand the mechanism by which cancer is caused, to identify by laboratory experiments those agents that might be expected, on the basis of that knowledge, to cause cancer in man, and then to eliminate them from the environment or, at least, to minimize the extent to which man comes into contact with them. With another, we look at the variation in the incidence of cancer in different groups of people, seek to characterize the conditions that give rise to exceptionally high risks, and then seek to modify those conditions.

The first strategy has had so many successes to its credit in recent years that we may reasonably hope that it will not be long before we come to understand what the characteristic features of a cancer cell are, the way such features are produced, and the conditions that enable clones of cells developed from them to multiply disproportionately and to spread throughout the body. Such knowledge should eventually indicate whether the disease can be eliminated altogether or whether, as most suspect, we shall have to lower our sights and aim only at reducing its incidence. Meanwhile, this approach has provided us with a variety of tests for detecting agents that can alter DNA, cause cells in culture to behave like the cells of a malignant tumour, or produce tumours in laboratory animals.

The results of such tests correlate well with human experience, in the sense that one or more of them is nearly always positive when an agent has been found to cause cancer in man. It should be noted, however, that when such an agent is identified, efforts are made to demonstrate its carcinogenicity in the laboratory that are far more intensive than those that are normally made in the testing of agents for which no human evidence is available, and some of the agents evaluated as confirmed or potential human carcinogens would not have been detected by the normal execution of routine tests.

There is, nevertheless, good reason to hope that the use of laboratory tests will draw attention to many of the avoidable causes of human cancer that are still unknown, including both external agents and those that are synthesized *in vivo,* and their use should prevent the introduction into industry of new powerful carcinogens with the consequent production of occupational hazards. Quantitative prediction is, however, still very uncertain, and it is not possible to tell from the results of tests on cells how readily, within two or three orders of magnitude, an agent will produce cancer in laboratory animals, nor from the effect of tests on one species how readily, to the same degree of accuracy, it will produce cancer in another. Nor can it yet be presumed that qualitative prediction from one species to another is always certain, when the mechanism is not known by which a particular type of tumour is produced in the laboratory, the tumour is produced in peculiarly susceptible species, or the dose required to produce the tumour is grossly large.

The second, epidemiological, or (as Richard Peto has called it) 'black box', strategy has hitherto obtained the most important practical results, in the sense that it has first drawn attention to most of the risks that have so far been recognized as causing large numbers of cancers throughout the world. It has the great merit that when it does reveal a cause it also provides an indication of the size of the effect; but its scope is limited by the variation in human behaviour and human exposure, and we can hope to obtain useful information from it only when that variation is both large and consistent. It is of value as a complement to the first mechanistic strategy for four principal purposes:

    (i)   to demonstrate risks that have been overlooked or suggested only tentatively by laboratory tests,

(ii) to estimate the level of exposure that produces the highest additional risk of disease that is likely to be socially acceptable (whatever society may determine that to be),
(iii) to ensure that disproportionate resources are not devoted to problems that are likely to be minor to the neglect of others that are likely to be more important, and
(iv) to check the correctness of conclusions about the causes of cancer by monitoring the effect of their removal.

In this Symposium, we are concerned only with the second of these uses of epidemiology and then only in certain defined circumstances: that is, when the validity of extrapolating the laboratory evidence to man is open to question and when epidemiological observations exist that do not clearly demonstrate the existence of a human hazard. It should be noted, however, that the selection of agents for discussion was made in 1982 and further evidence may have accumulated since then. The fact that an agent is included in the programme should not, therefore, be regarded as evidence that the participants, nor even the members of the organizing committee, necessarily believe that it is safe for man to use. Its inclusion is evidence only that, at the time the Symposium was planned, the majority of members of the organizing committee thought that the laboratory evidence was insufficient to justify the belief that the agent necessarily caused a material human hazard in the doses that men and women had habitually received and that the human evidence of carcinogenicity was negative or open to question. It is possible, therefore, that as our discussion proceeds, individuals, or the participants as a whole, may conclude that evidence of carcinogenicity of one or other substance is now so strong that a human hazard must be presumed to occur, if indeed it is not regarded as proved. Alternatively, of course, they may not; and the interest of our Symposium rests in the type of conclusion that, for practical purposes, can then be reached.

## ASSESSMENT OF LABORATORY EVIDENCE

As the programme has been arranged, we shall discuss first, for each agent, the laboratory evidence that suggests that it may present a carcinogenic hazard for man. Only very brief periods have been allowed for this part of our discussion, for two reasons: first, because it was thought that the results that had been obtained in the laboratory were reasonably clear and that, although their implications for man might be open to doubt, the facts were not; and, secondly, because most of the participants in the Symposium are epidemiologists or statisticians. It is hoped, therefore, that the laboratory evidence can be presented concisely and that it will not give rise to controversy.

Such laboratory evidence may be unequivocal or it may be complex and internally variable and it is difficult to classify it into categories with clearly different implications for extrapolation to man. The IARC has taken a lead in trying to define such categories and classified positive evidence for carcinogenicity as being either 'sufficient' or 'limited' (IARC, 1982); and this may be the best that can be done in our present state of ignorance. With the empiricism that toxicologists are accustomed to, chemicals with 'sufficient' evidence for carcinogenicity have generally been accepted as posing a potential hazard to man, even in the absence of detailed knowledge of the mechanism by which they exert their effect, while chemicals with 'limited' evidence are thought to require further study before any such conclusion can be drawn. More recently, however, it has come to be realized that carcinogens with 'sufficient' evidence for carcinogenicity may act by different mechanisms and that even for these substances reasons may be found that make extrapolation from one species to another inappropriate.

It seems, therefore, that experimental toxicological data on carcinogenesis can forewarn us of potential hazard to man, but that our final decision must rest on a full assessment of our total knowledge, including that derived from human observations and the economic, sociological and ethical features of society.

## ASSESSMENT OF EPIDEMIOLOGICAL EVIDENCE

*Limits of usefulness*

It is commonplace in logic that it is never possible to prove a hypothesis, only to disprove it, unless, of course, the hypothesis can be shown to be tautologous, as in mathematics. In practice, however, we are prepared to act as though biological hypotheses are true when they have been supported by experiment, and we confidently predict effects from our presumed knowledge of the mechanism by which they are produced and the way in which the agents act. We accept, too, for practical purposes, that epidemiological evidence can be strong enough, even when it is only circumstantial and no experiment has been carried out, to justify the conclusion that an agent is a cause of disease. If then, the agent is removed and the disease disappears (or, more likely, is reduced in incidence), few scientists think it worthwhile to argue whether or not a causal relationship has been established—unless, of course, some new evidence is obtained that casts doubt on the conclusion.

Whether epidemiological evidence, or any other type of evidence, can ever be said to show that an effect is *not* produced is not, I suggest, a very different matter. In theory, it cannot be done; in practice, we do it every day. Proof of absence of an effect by epidemiological means (in this practical sense) is, however, very much harder to achieve than proof that an effect *is* produced, particularly, perhaps, in the case of cancer, which has many causes and may not be produced by a particular agent in detectable amounts until several decades after exposure has occurred. Doll and Peto (1981) summarized their position as follows:

'Unless epidemiologists have studied reasonably large, well-defined groups of people who have been heavily exposed to a particular substance for two or three decades without apparent effect, they can offer no guarantee that continued exposure to moderate levels will, in the long run, be without material risk. For this reason prudent restrictions on occupational or public exposure to various substances often have to be based on indirect inference from laboratory studies of the agent being examined, without any direct evidence concerning its actual effect on humans. That is not to say that human evidence can ever be dispensed with. It is always relevant, but the weight that can be given to it varies greatly with the duration and intensity of the exposure experienced.... Negative human evidence may mean very little, unless it relates to prolonged and heavy exposure. If, however, it does and is consistent in a variety of studies (correlation studies over time, cohort studies of exposed individuals, and case-control studies of affected patients), whereas the laboratory evidence is limited in scope to, for instance, a particular type of tumour in a few species, negative human evidence may justify the conclusion that for practical purposes the agent need not be treated as a human carcinogen. In practice it is, of course, not usual for such perfect negative evidence to be available, but even less conclusive negative human evidence may help determine priorities between different lines of action.'

*Combination of data*

To this I should like to add, for the purpose of this Symposium, only three comments. First, I should have thought that the statement that human evidence is always relevant and can never be dispensed with, if available, was non-controversial, were it not that the Occupational Safety and Health Administration (1980) in the USA sought to lay down criteria for the admissibility of evidence that tended to show the lack of an effect. These criteria were that the data should refer to groups of subjects who had had at least 20 years' exposure, had been followed for at least 30 years, and were numerous enough for a 50% increase in the predicted type of cancer to be statistically significant. Such data would certainly carry substantial weight, but the exclusion of all other data would be unwise. A laboratory investigator can be advised to use so many animals, to test so many species, to treat at so many levels of dose, and to observe for a minimum length of time, as all these conditions are under his personal control—subject only to the constraints of finance and the availability of personnel. It is, however, unproductive to lay down similarly rigid rules for the epidemiologist, as experiments cannot be repeated and the conditions of the experiment are not under the epidemiologist's control. In practice, we can seek to overcome the deficiencies of one set of data only by combining it with sets from other sources, and any set that would be capable of showing a positive effect is worth considering. Indeed, it *must* be considered, if one is to avoid the trap described by Gaffey (personal communication), in which a positive effect is accepted because it appears in one set of data at the 1 in 20 level of significance, while 19 similar sets are exluded because they fail to satisfy more stringent requirements for the submission of negative results.

From what we know of the induction of cancer, it would be reasonable to require all sets of data to be set out in such a way as to show separately the results of observations made more than 10 years after first exposure, irrespective of whether this showed positive or negative results; but it would be unwise to exclude automatically even the first 10 years' data. In some circumstances, these could be crucial; as, for example, when examining the safety of an immunosuppressive drug that might replace azathioprine in the treatment of patients receiving renal transplants. The only safe rule is to consider the totality of the evidence, making sure, however, that it is set out in such a way that conclusions can be drawn about the presence or absence of effects for different durations of exposure, at different periods after exposure first began, and for different levels of dose.

*Negligible risks*

Secondly, I should like to echo the conclusions of a Royal Society (1983) Study Group that some induced events are 'of so low a frequency that the manager or regulator of risk can reasonably regard them as negligible in their overall impact on society, even though the consequences to the rare individual may be serious'. What this frequency should be is not easy to decide and almost certainly should be allowed to alter depending on the nature of the risk, its origin, and society's perception of it (which is, at least in part, subjective). It is, in any case, a matter for society itself to determine and not for the epidemiologist, whose job it is to calculate the risk and to explain what it means in practical terms. To help society reach a practical solution, the Study Group suggested that imposed risks can legitimately be treated as trivial at the point at which individuals, who are aware of the risks, would not commit significant resources of their own to reduce them. This, however, is a difficult point to assess, because so few people are conscious of the magnitude of small risks, and have little opportunity to demonstrate their preferences. To quote the Study Group's report again, 'There may be a wide spread of individuals' views... and decisions are likely to leave some people feeling they are exposed to risks calling for further control. There is a widely held view, though perhaps better described as speculation, that few people would commit their own resources to reduce an annual risk of death to themselves that was already as low as $10^{-5}$ and that even fewer would take action at an annual level of $10^{-6}$... while the manufacturer of a product might for the sake of his good name seek to keep the numbers of possible deaths very low, not all the inhabitants of the country will buy his product, and so... the figure of $10^{-6}$ is probably still appropriate, except perhaps if clear causal links are established in the risks from certain consumer products. In such circumstances we would consider $10^{-7}$ to be an annual level below which further control was certainly not justified, but even then the further problem of the very salient situation... may well remain.'

*Estimates of risk*

Thirdly, we need to consider whether we are going to pay more attention to the best estimate of a risk or to its upper or lower confidence limits. Both types of evidence are, of course, relevant, but we should, I suggest, give different emphasis to each depending on our prior hypothesis. When our data point to an unsuspected risk, we have all been accustomed to paying the greatest attention to the lower confidence limit to see whether the absence of an effect is plausible, and we shall want to pay similar attention to the upper confidence limit in the opposite situation in which the prior evidence suggests that a risk is likely to occur and we have failed to find it. When however, the prior evidence is inconclusive and has served only to stimulate enquiry, we should, I believe, make up our minds whether to postulate a risk on the totality of the evidence and then, if we do, put the greatest emphasis on the best estimate that we can make, modifying our confidence in it appropriately by knowledge of the width of its confidence limits.

*Categories of evidence*

The data that we shall be considering in the next three days will present many problems and will, I suspect, often provide exceptions to the general rules that I have tried to consider. They will not be solved by voting, but by the accumulation of experience and the conduct of experiments—if, indeed, they are ever solved at all. It would, however, help to decide what sort of evidence is needed in different situations if we were able to classify the existing human evidence (with no more than two or three dissentients) into one or other of the following five categories. They may not meet all eventualities, but they can be regarded as guides to the sort of conclusion that could be reached with respect to individual agents.

1. Evidence that supports the idea that the agent is carcinogenic to man
2. Evidence that is inadequate in quantity or quality to justify any useful conclusion
3. Evidence that is inadequate to permit a firm conclusion, but which suggests that the agent is unlikely to have produced a quantitatively large increase in risk under the conditions of exposure that have operated in the past
4. Evidence that weighs against the possibility that the agent is carcinogenic to man; but is not strong or consistent enough to outweigh laboratory evidence of carcinogenicity, even though the laboratory evidence is of doubtful generality
5. Evidence that weighs against the possibility that the agent is carcinogenic to man so strongly that one can either:
    (a) disregard, for practical purposes, laboratory evidence of doubtful generality; or
    (b) set a very low upper limit to the human risk produced by the conditions of exposure that have operated in the past if, on the basis of laboratory evidence, the agent must be presumed to be potentially carcinogenic to man.

If an agent is classed in category 3 or 4, it may be thought that the evidence is strong enough to justify waiting before any drastic action is taken, so long as arrangements can be made to obtain further human evidence that would ensure that even a small risk (should one exist) will eventually be detected.

The use of categories such as these has never been discussed by the International Agency for Research on Cancer and is not, of course, in any way a reflection of the Agency's policy. They represent a first attempt to deal with the problem of assessing the significance of 'negative' epidemiological evidence. They were found helpful by the participants in the discussion that followed, but remain the responsibility of the author.

## REFERENCES

Doll, R. (1983) Prospects for prevention. *Br. med. J.*, **286**, 445–453
Doll, R. & Peto, R. (1981) The causes of cancer: quantitative estimates of avoidable risks of cancer in the United States today. *J. natl Cancer Inst.*, **66**, 1193–1308

IARC (1982) *IARC Monographs on the Evaluation of the Carcinogenic Risk of Chemicals to Humans,* Suppl. 4, *Chemicals, Industrial Processes and Industries Associated with Cancer in Humans, IARC Monographs, Volumes 1 to 29,* Lyon

Occupational Safety and Health Administration (1980) Documentation of epidemiological studies. *Fed. Reg.,* **45,** Part B iv, Sections A, B, C, pp. 34–59

Royal Society (1983) *Risk Assessment: Report of a Study Group,* London

# STATISTICAL CONSIDERATIONS

# STATISTICAL CONSIDERATIONS

## N.E. DAY

*Unit of Biostatistics,
Division of Epidemiology and Biostatistics,
International Agency for Research on Cancer,
69372 Lyon Cedex 08, France*

In the *IARC Monographs* series (IARC, 1982), it is proposed that a substance for which there is sufficient evidence of carcinogenicity in experimental animals and inconclusive epidemiological data on its effect in man, should be treated as if it posed a carcinogenic risk to man. For a growing number of exposures, however, experimental data indicate the potential for a human risk, but the available epidemiological evidence does not demonstrate an effect. The present paper considers some of the statistical issues involved in drawing inferences from such apparently negative epidemiological data.

We shall start with some basic statistical concepts which, although trivial, put logical limits on the subsequent discussion. We consider the situation where a study results in an estimate of about unity for the relative risk associated with some exposure. When evaluating a relative risk distant from one, in terms of whether a causal relationship is a reasonable inference, three issues that need specific attention are whether the risk could have arisen by chance, by bias in the study design or due to the confounding effect of other factors (Hill, 1965). Similarly, in assessing whether a relative risk of one, or thereabouts, indicates no association with disease, chance, bias and confounding need to be excluded as reasonable alternative explanations.

*The role of chance*

Considering initially the simple comparison between two groups—one non-exposed, the other exposed—and an observed relative risk [or standardized mortality ratio (SMR), or odds ratio] of one, the role of chance can be examined by calculating the necessary confidence interval (95% or 99%, depending on the stringency required by the situation). The observed results could have arisen by chance for all values of the relative risk within this interval. For values outside the interval, chance would be considered an improbable explanation. In Table 1 we give confidence intervals for SMRs derived from cohort studies for various numbers of events of interest (e.g., lung cancer deaths), SMRs that are derived either from external comparisons, when the

Table 1. Confidence intervals for an estimated standardized mortality ratio of one, based on a cohort study

| Number of events in the exposed group | (a) Comparison with an external standard | | (b) Comparison with a non-exposed group of the same size as the exposed group | |
|---|---|---|---|---|
| | Lower | Upper | Lower | Upper |
| 100 | 0.801 | 1.216 | 0.750 | 1.334 |
| 200 | 0.866 | 1.149 | 0.816 | 1.225 |
| 1 000 | 0.939 | 1.064 | 0.915 | 1.094 |
| 5 000 | 0.972 | 1.028 | 0.959 | 1.042 |

non-exposed group is taken to be very large [Table 1(a)], or from internal comparisons, when the non-exposed group is taken to be the same size as the exposed group. In Table 2 confidence intervals are given for odds ratios derived from case-control studies, for different sample sizes and different proportions of exposure among the control group. If one takes 1000 events of interest as representing a large study—in fact, an order of magnitude larger than most industrial cohort studies—and 5000 as approaching the upper limit of studies that are likely to be done, then, ignoring all considerations other than chance, a negative study (i.e., one with a relative risk estimate of one) will usually be compatible with a 20% increase in risk and almost always be compatible with a 5% increase in risk.

For many of the industrial exposures that are suspected of being hazardous, on the basis usually of experimental results, studies of this size are unlikely to be possible, and one will be left, even if the estimated excess risk is zero, with confidence intervals containing risks of appreciable size (20 to 30%). Proposed legislation in some countries aims at reducing exposures to levels at which the excess risk, based on extrapolation procedures using data from animal carcinogenicity experiments, is of the order of $10^{-6}$. It is clear that epidemiology will never be able to determine whether such levels have been attained.

*Confounding effects and bias*

The role of unobserved confounding factors in weakening the conclusions of a study is often assumed to be confined to diminishing an observed positive effect. The effect can be the reverse, however. Observed relative risks of one can arise from genuine positive effects if there are unobserved variables that have a negative confounding effect. These variables would either be positively related to disease and negatively related to the exposure of interest, or the other way around. The extent to which the real relative risk differs from one, given an observed relative risk of one, is shown in Table 3, for a variety of situations (Breslow & Day, 1980).

A situation where such considerations might be of importance would be a study of lung cancer among some professional group, in which smoking histories were not obtainable (e.g., formaldehyde exposure of pathologists). An example where such considerations were of importance is the effect of breast cancer risk among women taking exogenous oestrogens for menopausal symptoms. When age at menopause was

Table 2. 95% confidence intervals for an estimated odds ratio of one, based on a case-control study (unmatched analysis, equal number of cases and controls)

| Proportion of controls exposed | Number of cases | | | |
|---|---|---|---|---|
| | 100 | 200 | 1 000 | 5 000 |
| 5% | 0.222–4.498 | 0.354–2.822 | 0.656–1.524 | 0.832–1.202 |
| 25% | 0.502–1.992 | 0.620–1.612 | 0.812–1.231 | 0.913–1.096 |
| 45% | 0.551–1.815 | 0.661–1.512 | 0.835–1.197 | 0.923–1.083 |

Table 3. Extent of negative confounding effect: ratio of the real relative risk associated with an exposure E to the estimated value if a confounding variable, C, is ignored[a]

(a) Odds ratio of C with disease = 2

| | | $P_2$ | | | |
|---|---|---|---|---|---|
| | | 0.1 | 0.3 | 0.5 | 0.8 |
| $P_1$ | 0.1 | 1.00 | 0.85 | 0.75 | 0.61 |
| | 0.3 | – | 1.00 | 0.87 | 0.72 |
| | 0.5 | – | – | 1.00 | 0.83 |
| | 0.8 | – | – | – | 1.00 |

(b) Odds ratio of C with disease = 5

| | | $P_2$ | | | |
|---|---|---|---|---|---|
| | | 0.1 | 0.3 | 0.5 | 0.8 |
| $P_1$ | 0.1 | 1.00 | 0.64 | 0.47 | 0.33 |
| | 0.3 | – | 1.00 | 0.73 | 0.52 |
| | 0.5 | – | – | 1.00 | 0.71 |
| | 0.8 | – | – | – | 1.00 |

[a] $P_1$, proportion among those exposed to E who are also exposed to C
$P_2$, proportion among those not exposed to E who are also exposed to C

ignored, a protective effect was observed, which disappeared after accounting for age at menopause and age at diagnosis, since oestrogen use is commoner among women with an early menopause (Casagrande *et al.*, 1976).

Similar to the effect of unobserved confounding variables, but arising from faults in study design rather than from the intrinsic nature of the factors, is the effect of bias, an issue of particular importance in case-control studies. Potential biases arising from the use of hospital controls have been well documented for tobacco smoking, and for saccharin consumption (Silverman *et al.*, 1983). In industrial cohort studies, the 'healthy worker effect' is a notorious source of bias. The effect is likely to be less for cancer than for other diseases and further reduced if cancer incidence rather than mortality is used as an end point. Nevertheless, for occupational cohort studies in

Fig. 1a. Cumulative risk of dying of mesothelioma in the absence of other causes of death among North American insulation workers first exposed to asbestos at age 15–24 (———), 25–34 (———), or over 35 (...), against age (upper graph) and years since first exposure (lower graph) (reproduced, with permission, from Peto et al., 1982)

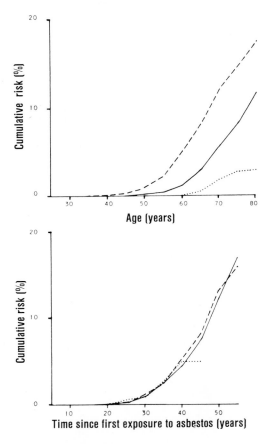

which the main weight of the results lies in the first 10 years of follow-up, relative risks in the region of one must be treated with caution.

*Dilution effects*

A major way in which results can suggest a more negative conclusion than is in fact justified is by including in the at-risk population individuals whose risk is likely to be low. This low risk could have arisen through insufficient exposure or through insufficient time elapsed since exposure began. We shall consider the time aspect first. In Figure 1 we give the evolution of risk with time since exposure started for a range of disease-exposure associations. Figure 1a indicates the evolution of risk of

Fig. 1b. Cumulative observed (. . .) and expected (———) probabilities of dying from lung cancer from 5 through 35 elapsed years since onset of work in an amosite asbestos factory, 1941–1945, for 188 men with two or more years of work experience. Observed lung cancer deaths shown are those classified according to the best evidence available. On the basis of death certificate information only, there were 34 lung cancer deaths (from Seidman et al., 1979).

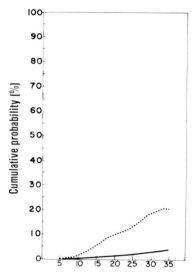

mesothelioma for those exposed to asbestos (Peto et al., 1982). There is very little indication of risk in the first 25 years after exposure starts.

Figure 1b gives the evolution of risk for lung cancer following short-term asbestos exposure, with no excess risk in the first 10 years, followed by a clear increase beginning in the 10–14-year period after exposure (Seidman et al., 1977).

Figure 1c shows the risk for heavily irradiated sites after radiotherapy for cancer of the cervix (Day & Boice, 1983); again, no risk is seen in the first 10 years, after which time the risk increases continually through the next 20 years of follow-up.

Figure 1d gives the risk for leukaemia following radiotherapy for ankylosing spondylitis, showing the typical wave pattern seen for leukaemia following irradiation (Smith & Doll, 1978).

Serious errors in interpretation can occur if the wrong time interval is used for assessing risk. An example is given in Table 4, of leukaemia following radiotherapy of cancer of the cervix uteri (Day & Boice, 1983). The small excess risk, which does not achieve statistical significance when all leukaemia is considered, conceals a fairly substantial excess if attention is confined to years 1–9 following irradiation, expecially if only the types of leukaemia known to be radiation-associated are included (underlining the importance of disease specificity).

Similar dilution of a real effect will occur if the dose levels included in the at-risk group are heavily weighted by low-effect dose levels. An example is provided by the

Fig. 1c. Observed to expected ratios of second cancer at sites at close and intermediate distance from the cervix, excluding uterine cancers, by time since diagnosis of cervical cancer for invasive cervical cancer patients treated with radiotherapy. 80% confidence intervals presented (from Day & Boice, 1983)

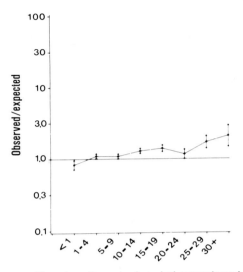

cervical cancer radiation study, in which no excess is seen for stomach cancer, with 86 observed and 86.1 expected 10 years or more after irradiation. The dose level received by the stomach would typically be in the range 1–2 Gy (100 to 200 rads) for Stage I and II cervical cancers, for patients who survive 10 years or more. In the study from Japan of the atomic bomb survivors (Kato & Schull, 1982), no risk for stomach cancer is seen among those exposed to less than 2 Gy. Above this dose, the risk increases steadily. The two sets of data are combined in Figure 2. If the data above 2 Gy were omitted, the dose-response relationship would appear essentially negative, with no indication of an upward tendency in the dose-response curve. In fact, up to 2 Gy the data are about as negative as one would ever expect to get from epidemiological studies. There are over 1000 cases with some exposure, and nearly 600 with exposure between 0.1 and 2 Gy. There is no sign of any increasing risk with increasing dose, and the information on radiation dose is sufficiently accurate for there to be little misclassification between the dose levels shown. As the data for above 2 Gy show, however, the stomach is certainly sensitive to the carcinogenic effects of radiation. All the data seem to be fitted reasonably well by a quadratic dose-response. This example, of course, makes no point other than the rather banal one that negative results refer only to the dose levels observed. Extrapolation to higher dose levels may be very misleading. The point, however, is important since in many industrial situations, for example, dose or exposure levels are not well documented. The

Fig. 1d. Observed and expected numbers of deaths from leukaemia according to time since first treatment (reproduced, with permission, from Smith & Doll, 1978).

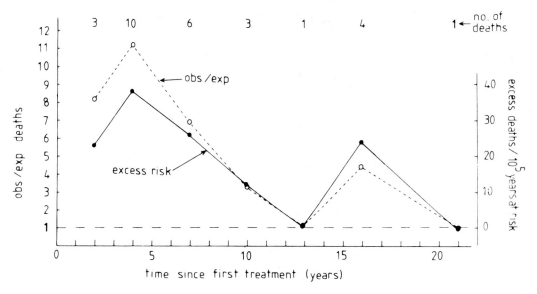

Table 4. Excess risk of leukaemia following radiation therapy for cancer of the cervix uteri

|  | Observed | Expected | Relative risk (with 95% confidence interval) |
|---|---|---|---|
| All leukaemia | 83 | 72.8 | 1.14 (0.91–1.41) |
| All leukaemia in years 1–9 following irradiation | 55 | 38.5 | 1.43 (1.08–1.86) |
| Acute and non-lymphocytic leukaemia in years 1–9 following irradiation | 45 | 24.2 | 1.86 (1.36–2.49) |

relevance of an apparently negative study to other situations would require careful consideration of what levels of exposure might have occurred.

*Dose-response relationships*

Dose-response relationships are often taken as strong evidence of causality, and the converse inference is sometimes made—that lack of a dose-response indicates no causal relationship. A good example is given by the link between coffee drinking and bladder cancer seen in some case-control studies, where the lack of dose-response, as exemplified by Table 5, led the authors to discount the possible causality of the relationship (Simon *et al.*, 1975). Lack of a dose-response relationship, however, must

Fig. 2. Relative risk for stomach cancer 10 years or more following radiation

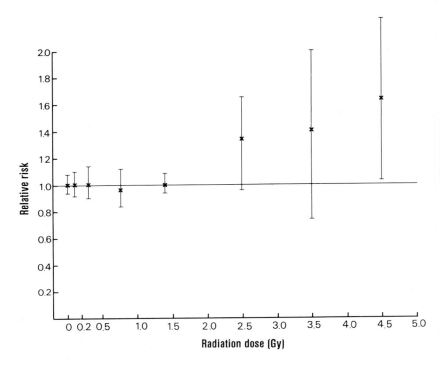

Table 5. Association between coffee drinking and tumours of the lower urinary tract

| No. of cups of coffee/day | Cases | Controls | Relative risk |
|---|---|---|---|
| <1 | 10 | 56 | 1.0 |
| 1–2 | 74 | 187 | 2.2 |
| 3–4 | 30 | 91 | 1.9 |
| 5+ | 20 | 48 | 2.3 |

From Simon et al. (1975)

be treated with care before being accepted as negative evidence. Two problems to be considered are the choice of exposure metameter to use, and the type of dose-response curve that may be biologically plausible. The choice of an exposure metameter needs to be guided both by the expected relationship between exposure and risk, in terms of dose level, time since first exposure, duration of exposure and so on, and by the accuracy with which the different aspects of exposure can be measured. In many industrial situations, very little is known of actual exposure levels, whereas the dates of starting and leaving employment are accurately recorded. If an attempt is made, within an exposed cohort, to relate exposure levels to risk, a completely negative

Table 6. Estimated asbestos exposure levels of men exposed between 1951 and 1955 (reproduced, with permission, from Peto, 1980)

| | Cumulative exposure (fibres/ml-years) to December 1971 | | | | | | |
|---|---|---|---|---|---|---|---|
| | 0– | 100– | 150– | 200– | 300– | 400– | Total |
| Men dying of lung cancer 20 years or more after first exposure | 1 | 1 | 0 | 4 | 1 | 1 | 8 |
| Other men[a] born 1901–1914 | 2 | 4 | 4 | 14 | 9 | 9 | 42 |

[a] For criteria of choosing controls, see text.

finding may emerge. An example is given in Table 6 (from Peto, 1980). In this example, the controls are chosen to have a similar age distribution to the lung cancer cases, and similar time since first exposure. The comparison is principally one of estimated dust levels. It would be surprising, however, if duration of exposure showed no relationship with risk, and often the use of cumulative exposure may be simply equivalent to duration, with reported exposure level being little more than a random effect.

In Table 7 we show the relationship between estimated mean bone-marrow dose (in Gy) and the risk for leukaemia among patients treated for ankylosing spondylitis. The leukaemia risk appears to be almost independent of dose. Two points need to be considered, however. First, only part of the active bone marrow is irradiated in the treatment of spondylitis, and for the irradiated marrow the radiation dose would be considerably higher than shown, perhaps twice as high. Second, cells receiving several Gy make be sterilized, and incapable of becoming leukaemic. One might therefore expect a dose-response of the form

$$\text{Excess risk}_\alpha \text{ dose} \cdot \exp(-\gamma \text{Dose})$$

for excess risk. Such a curve is shown in Figure 3, together with the data. The fit is reasonably good. Thus, in this example, an apparent lack of dose-response is consistent with a plausible biological model that would predict increasing risk with doses only at the low end of the dose range observed.

*Combining data from separate studies*

The considerations we have reviewed so far refer mainly to the results of a single study. One has also to consider inferences that might be drawn from a series of studies. Many of the issues remain the same, but some additional ones arise that are related to the consistency of the results, some of which derive from multi-comparison problems. Particularly in case-control studies, much more information is collected than is required to test a simple hypothesis, and a large number of comparisons can be made. Some will achieve a nominal statistical significance by chance. Further studies should then indicate which associations are real. In studies of genetic markers, especially of highly polymorphic loci such as the HLA system, it is now common

Table 7. Risk for leukaemia among ankylosing spondylitis patients given radiotherapy, in terms of estimated bone-marrow dose

| Mean marrow dose (Gy) | No. of deaths from leukaemia | No. of expected deaths | Relative risk |
|---|---|---|---|
| 0– | 2 | 0.54 | 3.7 |
| 1– | 7 | 0.86 | 8.1 |
| 2– | 3 | 0.97 | 3.1 |
| 3– | 4 | 1.12 | 3.6 |
| 4– | 3 | 0.99 | 3.0 |
| 5 | 6 | 0.92 | 6.5 |
| ≥6– | 3 | 0.46 | 6.5 |

Test of homogeneity of risk $\chi_6^2 = 4.6$; $p = 0.6$
Test for trend $\chi_1^2 = 0.025$

Fig. 3. Excess death rate from leukaemia more than 18 months (on average) after first treatment according to mean bone-marrow radiation dose. Curves are based on the following models: (0) ER = b; (1) ER = bD; (2) ER = $bDe^{-\lambda D}$; and (3) ER – $bD\,e^{-\lambda D}$, where ER is the excess leukaemia death rate, D is the mean bone-marrow dose, and b and λ are constants that were estimated for each model by the method of maximum likelihood (from Smith & Doll, 1982).

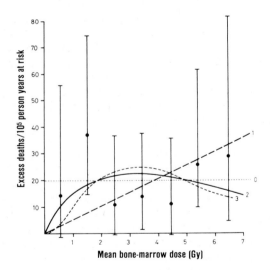

practice to take account explicitly of the multi-comparison problem. In other areas, it seems common practice to ignore it and, in the face of an overall negative result, to search for subgroups of the study population among whom an effect can be claimed. The question then is whether the same subgroups emerge in different studies. Saccharin, for example, may increase risk for bladder cancer among males who smoke in one study and among males who do not smoke in another study. This search for subgroups tends to generate spurious positive results.

Table 8. Increased risk for colon cancer following radiation exposure

| Study | Colon cancers | | |
|---|---|---|---|
| | Observed | Expected | SMR |
| Atomic bomb survivors[a] | 19 | 10.5 | 1.81 |
| Ankylosing spondylitis[b] | 16 | 10.4 | 1.54 |
| Metropathia haemorrhagica[c] | 21 | 13.5 | 1.56 |
| International Cervical Cancer Study[d] | 145 | 133.1 | 1.09 |

[a] Kato & Schull (1982)—5 years or more after exposure, exposure more than 1 Gy
[b] Smith & Doll (1982)—9 years or more after exposure, exposure estimated at 3 Gy
[c] Doll & Smith (1968)—5 years or more after exposure
[d] Day & Boice (1983)—10 years or more after exposure
Pooled estimate of SMR = 1.21 (1.04–1.41)
$\chi_1^2 = 2.52$; $p = 0.0059$ (one-sided)
Test of homogeneity $\chi_3^2 = 6.56$; $p = 0.087$

Multiple comparisons can, however, lead in a more subtle way to doubtful negative inferences. There may be, on occasion, many interrelated variables that an investigator might be tempted to incorporate into an analysis. Given a fairly weak effect for some exposure of interest, it may frequently be feasible to reduce the effect still further, to both statistically insignificant and inappreciable levels, by suitable choice of factors by which to stratify the data (Day et al., 1980). The question then is whether, in a series of studies, similar manipulations of the data lead to an overall negative conclusion, or whether a different approach is needed on each occasion. If the latter is the case, one must doubt the genuineness of the reported lack of effect. It is possible that the series of studies on reserpine might fall into this category.

Two examples of combining data may be of interest and relevance. First, data for cancer of the colon from the international cervical cancer radiation study: overall, among the irradiated group, 314 cases were observed with 302 expected, of which 145 occurred among 10-year survivors with 133.1 expected. The latter gives an SMR of 1.089, with a 95% confidence interval of 0.91–1.27. Under most circumstances, this would be regarded as a fairly strong negative result. The colon would normally receive a dose ranging from one or two to tens of Gy. The results of three other studies indicate that some risk for colon cancer is associated with not dissimilar radiation doses, as shown in Table 8. When the four studies are taken together, there is no strong evidence of heterogeneity and reasonable evidence of a positive effect (see Table 8). Excluding the ankylosing spondylitis series, among whom a higher risk of colon cancer might occur due to the increased prevalence of ulcerative colitis, makes little difference to these conclusions. Thus, relatively little positive evidence can outweigh, and yet be consistent with, a considerably larger body of apparently negative evidence.

The second example is perhaps that in which the smallest relative risk has been unquestionably established in cancer epidemiology, the association between gastric cancer and blood group A. A very large number of studies have been reported; Table 9 summarizes seven of the most important. The overall increase in risk is not high —some 22%—but the consistency of the seven studies is amazing (a $\chi^2$ for

Table 9. Association between stomach cancer and blood group A, from studies in different countries[a]

| Study | Cases | | Controls | | Relative risk |
|---|---|---|---|---|---|
| | % Type A | Number | % Type A | Number | |
| Aird et al. (1953) | 53 | 704 | 45 | 4 518 | 1.38 |
| Køster et al. (1955) | 51 | 413 | 44 | 14 304 | 1.33 |
| Aird et al. (1953) | 45 | 3 623 | 40 | 10 000 | 1.23 |
| Eisenberg et al. (1958) | 45 | 892 | 40 | 75 904 | 1.22 |
| Buckwalter et al. (1957) | 46 | 908 | 42 | 8 313 | 1.18 |
| Hogg & Pack (1957) | 44 | 237 | 41 | 12 917 | 1.13 |
| Billington & Sydney (1956) | 40 | 483 | 38 | 30 000 | 1.09 |

[a] From Nomura (1982)
Mantel-Haenszel estimate of relative risk = 1.226
95% confidence interval = 1.164–1.290
$\chi_1^2 > 60$
Test of homogeneity $\chi_6^2 = 5.13$

homogeneity of 5.1 on 6 degrees of freedom), given the numbers involved. This consistency gives great additional weight to the $\chi^2$ of over 50 obtained as an overall test of the effect. The consistency arises both from the precision and repeatability in determining blood groups and from the absence of confounding or biasing factors. Supposing that results of similar strength (summary relative risk = 1.22, summary $\chi^2 > 50$) were obtained in a different setting, with errors and inconsistencies in exposure measurement and varying degrees of bias and confounding occurring in the different studies. To achieve results that mimic this situation, we have arbitrarily increased the relative risk by 20% in three tables of Table 9 and reduced the relative risk by 20% in the other four tables. We show the result in Table 10. The figures in Table 10 represent the best one could hope to obtain from a series of different studies investigating some industrial exposure or dietary factor that is weakly associated with an increase in risk. The final interpretation is partly a matter of taste, but with four of the seven tables showing no effect, many epidemiologists would find the results at best inconclusive. Thus, by adding the degree of random variation one would expect when studying many occupational or lifestyle exposures to a clear-cut situation involving minor genetics risks, the power of the evidence to convince is substantially reduced, even if the formal statistical test remains the same.

## CONCLUSION

The considerations discussed above set clear limits on the degree of risk that epidemiology can aim at identifying. These levels are appreciably higher than the levels that it is customary to propose as representing 'acceptably low risk'. Thus, a 10% relative increase in risk for lung cancer represents an absolute increase in risk of 1%, which is orders of magnitude higher than the values of $10^{-5}$ or $10^{-6}$ often suggested in the context of low-dose extrapolation. To achieve levels at which risk should be acceptably low, two approaches might be taken. The first is to make use

Table 10. Random effects introduced into Table 9

| Study | Percentage of cases exposed | Relative risk | $\chi^2$ |
|---|---|---|---|
| 1 | 49 | 1.15 | 1.73 |
| 2 | 56 | 1.60 | 4.66 |
| 3 | 50 | 1.48 | 10.08 |
| 4 | 41 | 1.02 | 0.32 |
| 5 | 42 | 0.98 | −0.17 |
| 6 | 49 | 1.36 | 3.09 |
| 7 | 36 | 0.90 | −1.03 |

Mantel-Haenszel estimate of relative risk = 1.256
95% confidence interval = 1.194–1.320
$\chi_1^2 > 60$
Test of homogeneity $\chi_6^2 = 57.7$

of dose-response information. Thus, in a situation such as that shown in Figure 2, in which a quadratic dose-response curve fits the data well, it would seem reasonable to estimate risk at, say, 0.1 and 0.5 Gy, by means of a best-fitting quadratic dose-response curve, rather than to confine oneself to the data pertaining simply to 0.1 and 0.5 Gy. Such an approach depends on the availability of dose-response data, as, say, for asbestos (Acheson & Gardner, 1980), where the dose-response for lung cancer appears nearly linear. Often, such dose-response information will not be available. A second approach would be to use more sensitive markers of exposure-related effects — a field outside our present scope.

This paper has concentrated on small increases in relative risk for the more common cancers. If an exposure induces a type of cancer that is otherwise very rare, then one can clearly detect, epidemiologically, much smaller absolute increases in risk. However, exposures that cause rare cancers often increase risk for more common tumours as well, and it would seem perverse to base one's overall assessment of risk on the cancer that in absolute terms is responsible for less disease. Occurrence of a few cases of a rare cancer may be a useful pointer to the existence of a risk, but the absence of a rare cancer can hardly be taken as an indication of safety.

## REFERENCES

Acheson, E.D. & Gardner, M.J. (1980) *Asbestos: scientific basis for environmental control of fibres*. In: Wagner, J.R., ed., *Biological Effects of Mineral Fibres (IARC Scientific Publications No. 30)*, Vol. 2, Lyon, International Agency for Research on Cancer, pp. 737–754

Aird, I., Bentall, H.H. & Fraser Roberts, J.A. (1953) A relationship between cancer of stomach and the ABO blood groups. *Br. med. J., i,* 799–801

Billington, B.P. & Sydney, M.B. (1956) Gastric cancer. Relationships between ABO blood groups, site and research. *Lancet, ii,* 859–862

Breslow, N.E. & Day, N.E. (1980) *Statistical Methods in Cancer Epidemiology,*

Vol. 1, *The Analysis of Case-Control Studies (IARC Scientific Publications No. 32)*, Lyon International Agency for Research on Cancer

Buckwalter, J.A., Wohlwend, C.B., Colter, D.C., et al. (1957) The association of the ABO blood groups to gastric carcinoma. *Surg. Gynecol. Obstet.*, **104**, 176–179

Casagrande, J., Gerkins, V., Henderson, B.E., Mack, T. & Pike, M.C. (1976) Exogenous estrogens and breast cancer in women with natural menopause. *J. natl Cancer Inst.*, **56**, 839–841

Day, N.E. & Boice, J.D., Jr, eds (1983) *Second Cancer in Relation to Radiation Treatment for Cervical Cancer. Results of a Cancer Registry Collaboration (IARC Scientific Publications No. 52)*, Lyon, International Agency for Research on Cancer

Day, N.E., Byar, D.P. & Green, S.B. (1980) Overadjustment in case-control studies. *Am. J. Epidemiol.*, **112**, 696–706

Doll, R. & Smith, P.G. (1968) The long term effects of X irradiation in patients treated for metropathia haemorrhagica. *Br. J. Radiol.*, **41**, 362–368

Eisenberg, H., Greenberg, R.A. & Yesner, R. (1958) ABO blood group and gastric cancer. *J. chron. Dis.*, **8**, 342–348

Hill, A.B. (1965) The environment and health: association or causation. *Proc. R. Soc. Med.*, **58**, 295–300

Hogg, L. & Pack, G.T. (1957) The controversial relationship between blood group A and gastric cancer. *Gastroenterology*, **32**, 797–806

IARC (1982) *IARC Monographs on the Evaluation of the Carcinogenic Risk of Chemicals to Humans*, Suppl. 4, *Chemicals, Industrial Processes and Industries Associated with Cancer in Humans (IARC Monographs, Volumes 1 to 29)*, Lyon

Kato, H. & Schull, W.J. (1982) Studies of the mortality of A-bomb survivors. 7. Mortality, 1950–1978: Part I. Cancer mortality. *Radiat. Res.*, **90**, 395–432

Køster, K.H., Sindrup, E. & Seele, V. (1955) ABO blood groups and gastric acidity. *Lancet*, ***i***, 52–55

Nomura, A. (1982) Stomach. In: Schottenfeld, D. & Fraumeni, J.F., Jr, eds, *Cancer Epidemiology and Prevention*, Philadelphia, W.B. Saunders, pp. 624–637

Peto, J. (1980) Lung cancer mortality in relation to measured dust levels in an asbestos textile factory. In: Wagner, J.R., ed., *Biological Effects of Mineral Fibres (IARC Scientific Publications No. 30)*, Vol. 2, Lyon, International Agency for Research on Cancer, pp. 829–836

Peto, J., Seidman, H. & Selikoff, I.J. (1982) Mesothelioma mortality in asbestos workers: implications for models of carcinogenesis and risk assessment. *Br. J. Cancer*, **45**, 124–135

Seidman, H., Lilis, R. & Selikoff, I.J. (1977) Short term asbestos exposure and delayed cancer risk. In: Nieburgs, H.E., ed., *Prevention and Detection of Cancer, Part 1: Prevention*, Vol 1, *Etiology*, New York, Marcel Dekker, pp. 943–960

Seidman, H., Selikoff, I.J. & Hammond, E.C. (1979) Short-term absbestos work exposure and long-term observation. *Ann. N.Y. Acad. Sci.*, **330**, 61–89

Silverman, D.T., Hoover, R.N. & Swanson, G.M. (1983) Artificial sweeteners and lower urinary tract cancer: hospital vs. population controls. *Am. J. Epidemiol.*, **117**, 326–334

Simon, D., Yen, S. & Cole, P. (1975) Coffe drinking and cancer of the lower urinary tract *J. natl Cancer Inst.*, **54**, 587–593

Smith, P.G. & Doll, R. (1978) *Age- and time-dependent changes in the rates of radiation-induced cancers in patients with ankylosing spondylitis following a single course of X-ray treatment.* In: *Late Biological Effects of Ionizing Radiation,* Vol. 1, Vienna, International Atomic Energy Agency, pp. 205–218

Smith, P.G. & Doll, R. (1982) Mortality among patients with ankylosing spondylitis after a single treatment course with X rays. *Br. med. J.,* **284,** 449–460

# STATISTICAL CONSIDERATIONS:

# CONCLUSION

Rapporteur: P. Armitage

There was a very brief discussion in which the participants expressed admiration for Dr Day's survey. Two specific points were made, as follows:

(1) The 'healthy worker effect' is a complex phenomenon. In some studies, there may be an increase in incidence shortly after entry to a workforce, due to better diagnostic procedures. Mortality is likely to increase shortly after retirement, and active and retired employees should be grouped together. The 'healthy worker effect' may persist for more than a decade, in which case it is likely to be due to socio-economic confounding rather than to selection on the basis of health.

(2) When data from different studies are combined, confidence limits for the pooled estimate of relative risk should allow for possible heterogeneity between studies, e.g., by multiplying the standard error by the square root of the ratio of the chi-square heterogeneity statistic to its degrees of freedom.

# ORAL CONTRACEPTIVES AND BREAST CANCER

# ORAL CONTRACEPTIVES AND BREAST CANCER:

# LABORATORY EVIDENCE

### P. SHUBIK
*Green College, Oxford, UK*

It would not be possible to review all the available laboratory evidence on the carcinogenicity of all the oral contraceptives within the time allowed for this presentation. These data, are, in any case, readily available in Volume 21 of the *IARC Monographs* series (IARC, 1979). Many of the compounds involved are closely related both chemically and physiologically, and many overlapping effects are seen.

Studies with oral contraceptives in whole animals and in in-vitro systems pose a series of problems to the toxicologist that are of considerable general importance to the application of laboratory methods in many other areas.

Since oral contraceptives are composed largely of a mixture of oestrogens and progestational compounds, with some exceptions involving the use of single components, an immediate problem concerns the relevance of tests of the single components. Additionally, the question of the appropriateness of the test species plays a larger role in this instance than in many others. Finally, there is the confusion caused by the large number of commercial products now being marketed—perhaps at least 100. These different oral contraceptives bear many similarities to one another, but also have minor variations in chemical structure of the kind that often indicate to the experienced toxicologist the necessity for an additional set of studies.

For the purpose of this brief introduction, I have, therefore, selected a few, key, representative compounds as a basis for the overall discussion. The studies are summarized in Table 1.

I shall begin with the first oral contraceptive—Enovid—and its component parts, mestranol—the oestrogen and norethynodrel—the progestogen. The original oral contraceptive was composed of 1.5% mestranol and 98.5% norethynodrel. Mestranol when tested alone increased the incidences of pituitary and mammary tumours in mice, appeared to be inactive in rats, and increased the incidence of mammary tumours in dogs. When it was tested as Enovid—combined with norethynodrel—mice also developed squamous-cell tumours of the vagina and cervix, and rats developed mammary tumours and benign liver tumours.

Table 1. Carcinogenicity studies of oral contraceptives and their components

| Compound | Species | Tumour site | | | |
| --- | --- | --- | --- | --- | --- |
| | | Pituitary gland | Mammary gland | Liver | Genital organs |
| *Norethisterone* | | | | | |
| Alone | | | | | |
| | Mouse | + | − | + | +[a] |
| | Rat | + | − | + | − |
| | Dog | NA | NA | NA | NA |
| With ethinyloestradiol | | | | | |
| | Mouse | + | + | + | − |
| | Rat | − | + | + | − |
| *Medroxyprogesterone Acetate* (Depo-provera) | | | | | |
| Alone | | | | | |
| | Mouse | − | − | − | −[b] |
| | Rat | NA | NA | NA | NA |
| | Dog | − | + | − | − |
| *Chlormadinone acetate* | | | | | |
| Alone | | | | | |
| | Mouse | − | − | − | − |
| | Rat | − | − | − | − |
| | Dog | − | + | − | − |
| With mestranol | | | | | |
| | Mouse | + | − | − | − |
| | Rat | − | ± | − | − |
| | Dog | NA | NA | NA | NA |
| *Mestranol* | | | | | |
| Alone | | | | | |
| | Mouse | + | + | − | − |
| | Rat | − | − | − | − |
| | Dog | − | + | − | − |
| With norethynodrel | | | | | |
| | Mouse | + | +[c] | − | +[d] |
| | Rat | − | + | + | − |
| | Dog | − | + | − | − |
| *Norethynodrel* | | | | | |
| | Mouse | + | +[c] | − | − |
| | Rat | + | + | + | − |
| | Dog | − | − | − | − |

NA, not available
[a] Granulosa-cell tumours of the ovary
[b] After 20 months
[c] Castrated males
[d] Squamous-cell tumours of the vagina and cervix

The next representative compound I have selected is norethisterone, which is combined with ethinyloestradiol in several widely used oral contraceptives. When tested alone in mice, it is reported to enhance the incidences of pituitary and benign liver adenomas and of granulosa-cell tumours of the ovary. In rats, the incidences of pituitary and liver adenomas are increased. When tested in the combined form, mammary tumour incidence was enhanced in both mice and rats, liver tumours were noted in both species, and pituitary tumour incidence was enhanced in mice. No data on dogs are available.

Medroxyprogesterone acetate, or Depo-provera, is used alone as a contraceptive given every three months as a single intramuscular injection of 150 mg. No tumour was reported in mice; no study of rats is available, but mammary tumour incidence was enhanced in dogs.

Chlormadinone acetate is also used alone, at a dose of 0.5 mg/day. Studies in rats and mice were negative; mammary tumours were reported in dogs. When the compound was tested in combination with mestranol, pituitary tumour incidence was enhanced; controversial evidence of enhancement of mammary tumour incidence in mice was obtained. No studies of the combined preparation in dogs are available. In the USA neither Depo-provera nor chlormadinone is approved for use.

There is considerable debate over the significance of many of the tumours seen in these studies. Enhanced incidence of benign liver tumours in mice has been thought by many investigators to be irrelevant to human experience, but this conclusion may be contradicted in the case of oral contraceptives. The mammary tumours seen mainly in beagle dogs are mixed-cell tumours that bear little similarity to their human counterparts; again, their significance to human hazard has been questioned.

In summary, therefore, all the oral contraceptives and their component compounds are carcinogenic in rodents or dogs; the presently available data on monkeys are negative but not complete. Had these chemicals been destined for other uses, they would probably not have been permitted on the market. As matters stand, toxicologists are now provided with a unique opportunity for validating their approach to extrapolation of findings in animals to man.

## REFERENCE

IARC (1979) *IARC Monographs on the Evaluation of the Carcinogenic Risk of Chemicals to Humans,* Vol. 21, *Sex Hormones (II),* Lyon

# ORAL CONTRACEPTIVES AND BREAST CANCER:

# EPIDEMIOLOGICAL EVIDENCE

M.P. VESSEY

*Professor of Community Medicine, Oxford University, UK*

## INTRODUCTION

In most developed countries, breast cancer is the commonest malignant disease among women. In the UK alone, over 12 000 women die from the disease each year; it is estimated that 1 in 17 women in the UK eventually develop breast cancer, while in the USA the risk is even greater. Furthermore, in both of these, as well as in other industrialized countries, the incidence of the disease is increasing.

Available knowledge indicates clearly that both the etiology and the prognosis of breast cancer are influenced by hormonal factors, even though the precise mechanisms involved remain obscure (Kalache & Vessey, 1982). Accordingly, a proper assessment of the possible relationship between oral contraceptive use and breast cancer is of great public health importance, especially as over 50 million women world-wide are currently 'on the pill', while many millions more have used oral contraceptives in the past. Of course, the fact that breast cancer is to some degree dependent on hormonal factors does not imply that the pill will necessarily increase the risk; indeed it can be argued persuasively that the opposite might be true (Short & Drife, 1977).

A number of important general considerations need to be borne in mind when studying the literature on the relationship between oral contraceptives and breast cancer:

*Latent period*

There is usually an appreciable 'latent period' between first exposure to a carcinogen and the development of overt malignant disease. In addition, the cumulative effects of prolonged exposure or repeated exposure are likely to be of importance. Steroid contraceptives have been in widespread use only for about 20 years; relatively few women are likely to have had both the prolonged exposure and the extended period of follow-up required to evaluate carcinogenic effects with confidence. It should be noted, however, that if oral contraceptives were to alter the rate of growth of occult tumours, or were to affect the rate of change from a premalignant state to malignancy, an effect should now be apparent.

*Changes in preparation*

The types and doses of steroid in common use have changed markedly since the introduction of the pill. For example, (1) the dosages of both oestrogens and progestogens have been progressively reduced; (2) some new progestogens (e.g., norgestrel, desogestrel) have been introduced; (3) other progestogens (e.g., chlormadinone acetate, megestrol acetate) have been withdrawn; and (4) sequential preparations have largely disappeared, while triphasic and biphasic preparations have been developed. From this, it follows that even recent studies of the possible carcinogenic effects of oral contraceptives largely relate to discontinued products.

*Time of exposure*

Exposure to contraceptive steroids at particular times of life, such as during adolescence, during pregnancy, or during the perimenopausal years might be of special importance. In this context, it must be remembered that, over the years, women have tended to adopt the pill at younger and younger ages and to use hormonal contraception to delay the first birth as well as to space or limit later births. Again, during the last few years there has been a sharp fall in the numbers of women over 35 years of age who use the pill. It is important not to extrapolate the results of epidemiological studies to groups of women to whom they do not relate. For example, data derived from women who first started to use the pill in their mid-twenties may not be relevant to women who first started to use the pill in their mid-teens.

*High-risk groups*

It is possible that oral contraceptives might have no relationship with breast cancer in most women but, nonetheless, might have an effect on risk in those with predisposing factors such as a history of benign disease or a positive family history. To draw a well-known analogy from another field, it seems that the pill has an important effect on the risk of acute myocardial infarction only in the presence of other risk factors such as cigarette smoking.

## ORAL CONTRACEPTIVES AND BENIGN BREAST DISEASE

The use of oral contraceptives seems to be associated with a protective effect against clinically detectable benign breast disease. This topic has been reviewed in detail by Vessey (1983). The protective effect increases with duration of oral contraceptive use, as is clearly illustrated by the data from four major cohort studies shown in Table 1. In addition, the effect is probably confined to current or very recent users of oral contraceptives (Vessey, 1983); is probably attributable to the progestogen component

Table 1. Cohort studies of oral contraceptive use and benign breast disease

| Study | Type of disease[a] | No of cases | Relative risk by duration of oral contraceptive use (years) | | | | | |
|---|---|---|---|---|---|---|---|---|
| | | | 0 | <1 | 1– | 2– | 3– | 4– |
| Royal College of General Practitioners (1974) | Lumps, etc. (clinical diagnosis) | 859 | 1.0 | 0.9 | 0.9 | 0.8 | 0.6 | 0.5 |
| Eastern Massachusetts (Ory et al., 1976) | FC | 499 | 1.0 | 0.9 | 0.7 | ←—0.4—→ | | |
| | FA | 83 | 1.0 | ←—1.2—→ | | ←—0.5—→ | | |
| Oxford Family Planning Association (Brinton et al., 1981) | FC | 211 | 1.0 | ←—1.3—→ | | ←—0.9—→ | | 0.5 |
| | FA | 74 | 1.0 | ←—0.4—→ | | ←—0.5—→ | | 0.4 |
| | Lumps (not biopsied) | 311 | 1.0 | ←—0.8—→ | | ←—0.7—→ | | 0.5 |
| Walnut Creek (Ramcharan et al., 1981) | FC | 249 | 1.0 | ←—0.9—→ | | ←—0.8—→ | | 0.6 |
| | FA | 51 | 1.0 | ←—1.2—→ | | ←—0.8—→ | | 0.6 |

[a] FC, fibrocystic disease; FA, fibroadenoma

of the pill (Royal College of General Practitioners, 1977; Brinton et al., 1981); and may be restricted to the less serious forms of disease in which epithelial atypia are minimal or absent (Li Volsi et al., 1978). The last mentioned observation provides a possible explanation for the apparent paradox that oral contraceptives protect against benign breast disease, but not (as we shall see) against breast cancer.

## ORAL CONTRACEPTIVES AND BREAST CANCER

As oral contraceptives are used by more and more women, any major effect on the risk of breast cancer might be reflected in temporal trends in age-specific incidence rates or mortality rates. To the best of my knowledge, no change in trend has yet been reported that might be attributed to the influence of the pill. Mortality data for England and Wales are given in Table 2. As can be seen, the biggest increase in mortality has been in women aged 45–64 years, among whom the trend in rates has been steadily upward during the last 25 years.

Vital statistical data, however, have obvious limitations, and to make further progress we need to examine the results of analytical epidemiological studies. Many such studies have been published; they have been reviewed in detail by Kalache et al. (1983). Most of the early investigations included small numbers of subjects who had been exposed to oral contraceptives for short periods of time. Perhaps not surprisingly, the results were generally reassuring. Accordingly, in this review, I intend to confine myself for the most part to a discussion of five large case-control studies published during the last five years and to the latest reported figures for the major cohort studies.

Table 2. Breast cancer mortality, England & Wales 1955–1980

| Year | Rate per 100 000 population | | | | |
|---|---|---|---|---|---|
| | 15–24 years | 25–44 years | 45–64 years | 65–74 years | 75+ years |
| 1955 | 0.04 | 12 | 64 | 106 | 164 |
| 1960 | 0.20 | 12 | 66 | 105 | 161 |
| 1965 | 0.12 | 13 | 70 | 104 | 161 |
| 1970 | 0.23 | 14 | 78 | 109 | 158 |
| 1975 | 0.21 | 14 | 80 | 120 | 174 |
| 1980 | 0.05 | 13 | 83 | 122 | 183 |

*Case-control studies*

The main design features of the five studies under consideration are summarized in Table 3. The first two cover a wide span of ages and thus include some women who could have had little (or no) exposure to the pill. They also have a number of other limitations—in particular, the overall participation rate in the study by Paffenbarger *et al.* (1979) was only 68%, while the participation rate for breast cancer cases in the study by Brinton *et al.* (1982) was 86% but for controls was only 74%. Brinton *et al.* (1982) also rather arbitrarily excluded from their study patients who had had an artificial menopause.

*Total duration of oral contraceptive use.* This is, perhaps, the most important measure of exposure. The overall data from the five studies are summarized in Table 4. There is little evidence of any association between total duration of oral contraceptive use and breast cancer risk, although the data in the study by Paffenbarger *et al.* (1979) are slightly worrying. It should be noted that relatively few breast cancer cases have had very prolonged oral contraceptive use (say, 10 years or more).

*Interval since first oral contraceptive use.* Table 5 shows the overall data from the five studies. They are generally reassuring. The reports from the group at the Centers for Disease Control (1983) and from Rosenberg *et al.* (1984) include relative risks classified both by total duration of oral contraceptive use and by interval since first use. There is no indication in either study of any sinister pattern.

*Type of oral contraceptive used.* The published reports of three of the five studies include no data on type of oral contraceptive used. Brinton *et al.* (1982) provide a fairly detailed analysis by oestrogen content, with some suggestion of an elevated risk for users of high-dose preparations (100 µg or more). However, these authors found no 'distinct trend' when risk was related to cumulative lifetime oestrogen dose. Vessey *et al.* (1983) were unable to detect any association between breast cancer risk and oral contraceptive type, although their analysis was limited to the proportions of breast cancer cases and controls who had ever used particular individual preparations or groups of preparations. It is clear that work on the relationship between pill type and breast cancer is handicapped not only by the inadequate recall of brand names by study participants but also by the lack of any satisfactory way of grouping different oral contraceptives, taking into account both the oestrogen and the progestogen components.

Table 3. Main features of major case-control studies

| Reference | Place | Age range (yrs) | Time cases diagnosed | No. of cases | Nature of cases | No. of controls | Nature of controls | Matching criteria | Method of data collection |
|---|---|---|---|---|---|---|---|---|---|
| Paffenbarger et al. (1979) | San Francisco, USA | All ages | 1973–1977 | 1432 | Newly diagnosed at participating hospitals | 2560 | Hospital patients | Age, race, hospital, admission date | Home interview |
| Brinton et al. (1982) | Multi-centre, USA | All ages from 35 up | 1973–1977 | 963 | Newly diagnosed in BCDDP[a]. Those with artificial menopause excluded. Whites only. | 858 | Screening participants | Age, centre, time of entry to project, continuation in project | Home interview by trained nurses |
| Centers for Disease Control (1983) | Multi-centre, USA | 20–54 | 1980–1981 | 689 | Newly diagnosed - identified by cancer registries | 1077 | Population-based | Not individually matched. Selected in same age group within same geographic area. | Home interview |
| Vessey et al. (1983) | London & Oxford, UK | 16–50 | 1968–1980 | 1176 | Newly diagnosed at nine hospitals. Married only. | 1176 | Hospital patients | Age, parity, hospital admission date | Interviewed in hospital by trained nurse or social worker |
| Rosenberg et al. (1984) | Multi-centre, USA & Canada | 20–59 | 1976–1981 | 1191 | Newly diagnosed at participating hospitals | 5026 | Hospital patients | Not individually matched. Selected in same age group within same hospital | Interviewed in hospital by trained nurse |

[a] BCDDP, Breast Cancer Detection Demonstration Project

Table 4. Breast cancer risk in case-control studies in relation to total duration of oral contraceptive use (the listing of 19 years as the upper limit of use is arbitrary)

| Total duration of oral contraceptive use (yrs) | Paffenbarger et al. (1979) | | Brinton et al. (1982) | | Centers for Disease Control (1983) | | Vessey et al. (1983) | | Rosenberg et al. (1984) | |
|---|---|---|---|---|---|---|---|---|---|---|
| | Relative risk | No. of cases | Relative risk | No. of cases | Relative risk | No. of cases | Relative risk | No. of cases | Relative risk | No. of cases |
| 0 | 1.0 | 1 106 | 1.0 | 738 | 1.0 | 294 | 1.0 | 639 | 1.0 | 794 |
| 1 | 1.1 | 112 | | | 0.9 | 147 | 0.9 | 203 | 0.9 | 97 |
| 2 | 1.2 | 39 | 0.9 | 113 | | | | | | |
| 3 | 1.4 | 49 | | | 1.2 | 125 | 1.0 | 145 | 1.0 | 127 |
| 4 | | | | | | | | | | |
| 5 | 1.3 | 33 | 1.2 | 36 | | | | | | |
| 6 | | | | | 1.0 | 36 | 1.2 | 123 | 1.3 | 88 |
| 7 | | | | | | | | | | |
| 8 | | | 1.5 | 35 | | | | | | |
| 9 | | | | | 0.7 | 29 | | | | |
| 10 | | | | | | | | | | |
| 11 | | | | | | | | | | |
| 12 | | | | | | | | | | |
| 13 | 1.4 | 72 | | | | | | | | |
| 14 | | | 1.0 | 29 | | | 1.0 | 66 | 0.8 | 25 |
| 15 | | | | | 0.9 | 35 | | | | |
| 16 | | | | | | | | | | |
| 17 | | | | | | | | | | |
| 18 | | | | | | | | | | |
| 19 | | | | | | | | | | |

*Oral contraceptive use at different times of life.* Analyses of the relationship between oral contraceptive use and breast cancer risk in subgroups defined by age and parity were carried out in four of the studies and by menopausal status in all five. In general, the results were unrevealing, although Brinton et al. (1982) found that premenopausal women who used the pill after the age of 40 had an approximately 50% increase in risk and suggested that this might be due to 'artificial prolongation of a premenopausal rate of disease incidence'. It is of interest that Jick et al. (1980), in a much smaller study, had previously reported a positive association between current oral contraceptive use and breast cancer risk in premenopausal women over 45 years of age. Vessey et al. (1983) also found some elevation of risk in women aged 46–50 years (but not in those aged 41–45 years), but this was not confined to current users, nor was it influenced by menopausal status.

Anxiety about a possible harmful effect of prolonged oral contraceptive use before first term pregnancy followed publication of the data shown in Table 6 by Pike et al. (1981). The study involved 163 women in Los Angeles County in whom breast cancer had been diagnosed at age 32 years or less. Relevant findings in the five case-control studies under review are summarized in Table 7. The data are broadly reassuring, but the numbers of cases and controls with prolonged oral contraceptive use before first term pregnancy are small, and few of the cases developed cancer at as young an age as those in the study by Pike et al. (1981). Clearly, more information is required about

Table 5. Breast cancer risk in case-control studies in relation to interval since first oral contraceptive use (the listing of 19 years as the upper limit of use is arbitrary)

| Interval since first oral contraceptive use (yrs) | Paffenbarger et al. (1979) | | Brinton et al. (1982) | | Centers for Disease Control (1983) | | Vessey et al. (1983) | | Rosenberg et al. (1984) | |
|---|---|---|---|---|---|---|---|---|---|---|
| | Relative risk | No. of cases | Relative risk | No. of cases | Relative risk | No. of cases | Relative risk | No. of cases | Relative risk | No. of cases |
| 0 | 1.0 | 1 106 | 1.0 | 738 | 1.0 | 294 | 1.0 | 639 | 1.0 | 794 |
| 1 | | | | | | | | | | |
| 2 | | | | | | | 0.8 | 100 | 0.6 | 16 |
| 3 | 1.2 | 105 | 0.9 | 33 | | | | | | |
| 4 | | | | | | | | | | |
| 5 | | | | | 1.4 | NS | | | | |
| 6 | | | | | | | 1.2 | 172 | | |
| 7 | | | | | | | | | 1.0 | 78 |
| 8 | | | | | | | | | | |
| 9 | 1.3 | 94 | 1.2 | 124 | | | | | | |
| 10 | | | | | | | 1.2 | 153 | | |
| 11 | | | | | 1.2 | NS | | | | |
| 12 | | | | | | | | | 1.0 | 143 |
| 13 | | | | | | | | | | |
| 14 | | | | | 1.0 | NS | | | | |
| 15 | 1.2 | 94 | 0.8 | 56 | | | 0.7 | 112 | | |
| 16 | | | | | | | | | | |
| 17 | | | | | 0.9 | NS | | | 1.1 | 101 |
| 18 | | | | | | | | | | |
| 19 | | | | | | | | | | |

NS, not stated

Table 6. Risk of breast cancer in relation to use of oral contraceptives before first full-term pregnancy (modified from Pike et al., 1981)

| Duration of oral contraceptive use (months) | Cases | | Controls | | Relative risk |
|---|---|---|---|---|---|
| | No. | % | No. | % | |
| 0 | 79 | 48.5 | 141 | 52.2 | 1.0 |
| 1–48 | 53 | 32.5 | 103 | 38.2 | 1.0 |
| 49–96 | 24 | 14.7 | 22 | 8.1 | 2.2 |
| 97– | 7 | 4.3 | 4 | 1.5 | 3.5 |
| Total | 163 | 100.0 | 270 | 100.0 | |

Test for linear trend, $p = 0.009$

this extremely important question; it is likely to be obtained from the study by the Centers for Disease Control, which, when completed, should include about 4000 cases of breast cancer.

*Oral contraceptive use in high-risk groups.* Most attention has been concentrated on women with a family history of breast cancer or with a history of benign breast disease. Brinton et al. (1982) reported rather worrying findings for oral contraceptive

Table 7. Risk of breast cancer in relation to use of oral contraceptives before first term pregnancy[a]

| Study | Duration of oral contraceptive use (months) | Cases No. | % | Controls No. | % | Relative risk | |
|---|---|---|---|---|---|---|---|
| Paffenbarger et al. (1979); Paffenbarger et al. (1980) (Premenopausal women only. Risks adjusted for age, race, age at first pregnancy) | 0 | 574 | 94.9 | 1 146 | 97.9 | 1.0 | |
| | 1–17 | 13 | 2.1 | 9 | 0.8 | 2.6 | } $p < 0.05$ |
| | 18– | 18 | 3.0 | 16 | 1.4 | 2.6 | |
| | Total | 605 | 100.0 | 1 171 | 100.0 | | |
| Centers for Disease Control (1983) (Parous women only. Risks adjusted for menopausal status and age at first pregnancy)[b] | 0 | 323 | 83.5 | 542 | 84.6 | 1.0 | |
| | 1–35 | 45 | 11.6 | 57 | 8.9 | 1.3 | NS |
| | 36– | 19 | 4.9 | 42 | 6.5 | 1.3 | |
| | Total | 387 | 100.0 | 641 | 100.0 | | |
| Vessey et al. (1982); Vessey et al. (1983) (Risks adjusted for social class, age at menarche, age at first pregnancy, menopausal status, smoking, history of breast biopsy, family history of breast cancer) | 0 | 1 072 | 91.2 | 1 065 | 90.6 | 1.0 | |
| | 1–12 | 43 | 3.7 | 50 | 4.3 | 0.7 | NS |
| | 13–48 | 37 | 3.1 | 37 | 3.1 | 0.8 | |
| | 49– | 24 | 2.0 | 24 | 2.0 | 0.8 | |
| | Total | 1 176 | 100.0 | 1 176 | 100.0 | | |
| Rosenberg et al. (1984) (Parous women only. Risks adjusted for age and age at first pregnancy. Note that controls much younger than cases)[c] | 0 | 643 | 93.5 | 1 946 | 84.0 | 1.0 | |
| | 1–11 | 14 | 2.0 | 149 | 6.4 | 0.8 | NS |
| | 12–35 | 21 | 3.1 | 133 | 5.7 | 1.3 | |
| | 36– | 10 | 1.4 | 91 | 3.9 | 0.9 | |
| | Total | 688 | 100.0 | 2 319 | 100.0 | | |

NS, Not significant
[a] The study by Brinton et al. (1982) included no detailed information on oral contraceptive use before first pregnancy
[b] Also excluded were 136 cases and 303 controls who were in strata of menopausal status and age at first term pregnancy in which no women used oral contraceptives before first term pregnancy.
[c] Also excluded were 291 parous cases and 1661 parous controls who first used oral contraceptives after the first term pregnancy, or whose time of first use was unknown, or whose age at first pregnancy was unknown.

users in both of these high-risk groups, although their results were not statistically significant. A similar elevation in risk (also not statistically significant) was described by Paffenbarger et al. (1979) in women with a maternal history of breast cancer, but not in those with a history of benign breast disease. Negative results for both high-risk groups were reported in the studies from the Centers for Disease Control (1983), Vessey et al. (1983) and Rosenberg et al. (1984).

*Cohort studies*

Data on oral contraceptive use and breast cancer have been reported from four large cohort studies: (1) the Royal College of General Practitioners (1981) study, which includes 46 000 UK women recruited by 1400 family doctors; (2) the Oxford Family Planning Association Contraceptive study (Vessey et al., 1981), which includes 17 000 UK women recruited at 17 family planning centres; (3) the Walnut Creek Contraceptive Drug Study (Ramcharan et al., 1981), which includes 16 500 North American women who joined the study by having a general health check at an automated multitest laboratory; and (4) a study conducted in eastern Massachussets (Trapido, 1981), which includes 96 000 women who were identified from residence lists and mailed a questionnaire in 1970. Findings in relation to total duration of oral contraceptive use are summarized in Table 8. The data are clearly reassuring as far as they go, but the numbers of cases involved are rather small.

In the Royal College of General Practitioners Study, there was some evidence that breast cancer risk might be increased in oral contraceptive users under 35 years of age (relative risk of 'ever' users to 'never' users, 2.8 to 1; $p = 0.05$). There was no indication of any such effect in the other three studies. Trapido (1981) found the relative risk associated with oral contraceptive use in nulliparous women to be 2.1, but the figure was not statistically significantly different from unity.

The Royal College of General Practitioners Study and the Oxford Family Planning Association Contraceptive Study, both of which are continuing, will eventually yield a very large amount of information about the risk of breast cancer in oral contraceptive users. Neither study will, however, be able to provide much data about women starting to use the pill at a very young age, while much of the exposure information will inevitably relate to pills that are no longer used.

*Oral contraceptive use and prognosis of breast cancer*

Spencer et al. (1978) compared 44 newly diagnosed breast cancer patients with recent oral contraceptive use with 44 other newly diagnosed breast cancer patients who were non-users of the pill and found prognosis to be better in the former group. Similar findings were later reported by the same workers (Mathews et al., 1981) in an investigation involving an additional 93 breast cancer patients who had been taking the pill and 93 'controls' studied in an identical way.

Vessey et al. (1983) have reported on clinical stage of breast cancer at diagnosis in 572 breast cancer patients included in their case-control study. As is shown in Table 9, women who had never used oral contraceptives presented with more advanced

Table 8. Breast cancer risk in cohort studies in relation to total duration of oral contraceptive use (the listing of 19 years as the upper limit of use is arbitrary)

| Total duration of oral contraceptive use (yrs) | Royal College of General Practitioners (1981) | | Vessey et al. (1981) (Oxford Family Planning Association Study) | | Ramcharan et al. (1981) (Walnut Creek Study) | | Trapido (1981) (Eastern Mass. Study) | |
|---|---|---|---|---|---|---|---|---|
| | Relative risk | No. of cases[a] | Relative risk | No. of cases | Relative risk | No. of cases | Relative risk | No. of cases |
| 0  | 1.0 | 58 | 1.0 | 33 | 1.0 | 64 | 1.0 | 370 |
| 1  | 0.7 | 2  |     |    |     |    | 1.1 | 34  |
| 2  | 1.3 | 4  | 0.7 | 10 |     |    |     |     |
| 3  | 1.5 | 5  |     |    |     |    | 0.6 | 17  |
| 4  | 0.6 | 2  |     |    |     |    |     |     |
| 5  | 2.1 | 7  |     |    |     |    | 0.8 | 19  |
| 6  | 0.9 | 3  | 1.2 | 21 |     |    |     |     |
| 7  | 1.4 | 4  |     |    |     |    |     |     |
| 8  | 0.8 | 2  |     |    |     |    |     |     |
| 9  |     |    |     |    |     |    |     |     |
| 10 |     |    |     |    | 1.2 | 67 |     |     |
| 11 |     |    |     |    |     |    |     |     |
| 12 |     |    |     |    |     |    | 0.9 | 15  |
| 13 |     |    |     |    |     |    |     |     |
| 14 | 1.1 | 7  | 0.9 | 8  |     |    |     |     |
| 15 |     |    |     |    |     |    |     |     |
| 16 |     |    |     |    |     |    |     |     |
| 17 |     |    |     |    |     |    |     |     |
| 18 |     |    |     |    |     |    |     |     |
| 19 |     |    |     |    |     |    |     |     |

[a] Current users only

tumours than those who had used them during the year before detection of cancer, while past users were in an intermediate position. These differences in staging were reflected in the pattern of survival.

Possible explanations for these observations include 'surveillance bias' among oral contraceptive users, leading to earlier diagnosis [although Vessey et al. (1979) found little evidence to support this suggestion], or a beneficial effect of oral contraceptives on tumour growth and spread.

## CONCLUSIONS

The available data about a possible association between oral contraceptive use and the risk of developing breast cancer are reassuring. For the reasons discussed in the introduction to this paper, however, more information is needed before definite conclusions can be drawn. This is especially true in relation to certain subgroups of women who may be specially vulnerable to any adverse effect, in particular:
—users with prolonged exposure at a young age
—users with prolonged exposure before first term pregnancy

Table 9. Stage classification of 572 patients with breast cancer (modified from Vessey et al., 1983)

| Clinical stage | Use of oral contraceptives[a] | | | | | |
|---|---|---|---|---|---|---|
| | Never used | | Used only in past | | Used recently | |
| | No. | % | No. | % | No. | % |
| I | 196 | 55.4 | 83 | 64.3 | 66 | 74.2 |
| II | 74 | 20.9 | 23 | 17.8 | 11 | 12.4 |
| III–IV | 84 | 23.7 | 23 | 17.8 | 12 | 13.5 |
| Total | 354 | 100.0 | 129 | 100.0 | 89 | 100.0 |

[a] 'Used recently' indicates use during year before detection of lump. 'Used only in past' indicates use only before that time.
$\chi^2_{(4)} = 11.75$; $p = 0.02$

— nulliparous users
— users with a history of benign breast disease
— users with a positive family history
— users with exposure during the perimenopausal years.

Very few data are available at present on breast cancer risk in relation to particular oral contraceptive formulations.

Women who have never used oral contraceptives seem to present with more clinically advanced tumours than those who have used them. The differences in staging are reflected in patterns of survival. This difference may represent either surveillance bias or a biological effect of the pill.

Prolonged oral contraceptive use protects against benign breast disease, although the effect may be confined to lesions without malignant potential.

## REFERENCES

Brinton, L.A., Vessey, M.P., Flavel, R. & Yeates, D. (1981) Risk factors for benign breast disease. *Am. J. Epidemiol.*, **113**, 203–214

Brinton, L.A., Hoover, R., Szklo, M. & Fraumeni, J.F. (1982) Oral contraceptives and breast cancer. *Int. J. Epidemiol.*, **11**, 316–322

Centers for Disease Control (1983) Long-term oral contraceptive use and the risk of breast cancer. *J. Am. med. Assoc.*, **249**, 1591–1595

Jick, H., Walker, A.M., Watkins, R.N., D'Ewart, D.C., Hunter, J.R., Danford, A., Madsen, S., Dinan, B.J. & Rothman, K.J. (1980) Oral contraceptives and breast cancer. *Am. J. Epidemiol.*, **112**, 577–585

Kalache, A. & Vessey, M.P. (1982) Risk factors for breast cancer. *Clin. Oncol.*, **1**, 661–678

Kalache, A., McPherson, K., Barltrop, K. & Vessey, M.P. (1983) Oral contraceptives and breast cancer. *Br. J. Hosp. Med.*, **30**, 278–283

Li Volsi, V.A., Stadel, B.V., Kelsey, J.L., Holford, T.R. & White, C.W. (1978) Fibrocystic disease in oral contraceptive users. A histopathological evaluation of epithelial atypia. *New Engl. J. Med.*, **299**, 381–385

Mathews, P.N., Millis, R.R. & Hayward, J.L. (1981) Breast cancer in women who have taken contraceptive steroids. *Br. med. J., 282,* 774–776

Ory, H., Cole, P., MacMahon, B. & Hoover, R. (1976) Oral contraceptives and reduced risk of benign breast diseases. *New Engl. J. Med., 294,* 419–422

Paffenbarger, R.S., Kampert, J.B. & Chang, H-G. (1979) Oral contraceptives and breast cancer risk. *Editions de l'INSERM, 83,* 93–114

Paffenbarger, R.S., Kampert, J.B. & Chang, H-G. (1980) Characteristics that predict risk of breast cancer before and after the menopause. *Am. J. Epidemiol., 112,* 258–268

Pike, M.C., Henderson, B.E., Casagrande, J.T., Rosario, I. & Gray, G.E. (1981) Oral contraceptive use and early abortion as risk factors for breast cancer in young women. *Br. J. Cancer, 43,* 72–76

Ramcharan, S., Pellegrin, F.A., Ray, R. & Hsu, J-P. (1981) *The Walnut Creek Contraceptive Drug Study,* Vol. III, Washington DC, US Government Printing Office

Rosenberg, L., Miller, D.R., Kaufman, D.W., Helmrich, S.P., Stolley, P.D., Schottenfeld, D. & Shapiro, S. (1984) Breast cancer and oral contraceptive use. *Am. J. Epidemiol., 119,* 167–176

Royal College of General Practitioners (1974) *Oral Contraceptives and Health,* London, Pitman Medical

Royal College of General Practitioners (1977) Effect on hypertension and benign breast disease of progestogen component in combined oral contraceptives. *Lancet, i,* 624

Royal College of General Practitioners (1981) Breast cancer and oral contraceptives: findings in Royal College of General Practitioners Study. *Br. med. J., 282,* 2089–2093

Short, R.V. & Drife, J.O. (1977) The aetiology of mammary cancer in man and animals. *Symp. Zool. Soc. Lond., 41,* 211–230

Spencer, J.D., Millis, R.R. & Hayward, J.L. (1978) Contraceptive steroids and breast cancer. *Br. med. J., i,* 1024–1026

Trapido, E.J. (1981) A prospective cohort study of oral contraceptives and breast cancer. *J. natl Cancer Inst., 67,* 1011–1015

Vessey, M.P. (1983) *Contraception and benign breast disease.* In: Renaud, R. & Gairard, B., eds, *Proceedings of V<sup>es</sup> Journées de la Société Française de Sénologie et de Pathologie Mammaire,* Paris, Masson, pp. 63–69

Vessey, M.P., Doll, R., Jones, K., McPherson, K. & Yeates, D. (1979) An epidemiological study of oral contraceptives and breast cancer. *Br. med. J., i,* 1755–1758

Vessey, M.P., McPherson, K. & Doll, R. (1981) Breast cancer and oral contraceptives: findings in the Oxford-Family Planning Association Contraceptive Study. *Br. med. J., 287,* 2093–2094

Vessey, M.P., McPherson, K., Yeates, D. & Doll, R. (1982) Oral contraceptive use and abortion before first term pregnancy in relation to breast cancer risk. *Br. J. Cancer, 45,* 327–331

Vessey, M., Baron, J., Doll, R., McPherson, K. & Yeates, D. (1983) Oral contraceptives and breast cancer: Final report of an epidemiological study. *Br. J. Cancer, 47,* 455–462

# ORAL CONTRACEPTIVES AND BREAST CANCER:

## CONCLUSION

### Rapporteur: B. MACMAHON

Dr Vessey's paper pointed out that breast cancer patients who have never used oral contraceptives tend to present with tumours at a later stage than those seen in oral contraceptive users. In the discussion, it was pointed out that this difference is reflected in lower case fatality rates from breast cancer for oral contraceptive users, a fact which might introduce methodological difficulties, particularly, for example, in comparing incidence and prevalence series in case-control studies.

Questions were raised about the relationship of level of oestrogen exposure of a woman on the pill to that of one who is not. This question is not clearly answered, but it probably varies greatly from one preparation to another. It was pointed out that there is no published information on the oestrogen-receptor status of tumours following use of oral contraceptives.

Additional data were discussed with regard to the question of whether there is evidence of increased risk of breast cancer associated with oral contraceptive use at the extremes of reproductive life. The actual data on this are conflicting. The 'oestrogen window' hypothesis of Korenman would seem to predict that breast cancer incidence should be reduced by use of oral contraceptives around the time of menarche and menopause, but there is no evidence for this.

It was pointed out that different studies have frequently found elevated breast cancer risks in certain sub-groups of users, but that these are seldom the same sub-groups. With sufficiently searching analysis one can always find such groups, and one should be distrustful of them unless they are large or turn up repeatedly in several studies.

The relationship of oral contraceptive use to breast cancer is a classic problem in which laboratory and human evidence do not synthesize: entirely different inferences are reached according to which body of data—human or animal—is reviewed.

In spite of the reasonable degree of assurance provided by the data produced to date, this is no time to discontinue studies of effects of oral contraceptives on the breast. In particular, further information is needed on the six groups of women identified in Dr Vessey's paper:

—users with prolonged exposure at a young age
—users with prolonged exposure before first term pregnancy

—nulliparous users
—users with a history of benign breast disease
—users with a positive family history
—users with exposure during the perimenopausal years.

Discussion of oral contraceptives was limited to consideration of their possible effects on breast cancer risk. At present, the epidemiological evidence weighs against the possibility that the compounds are mammary carcinogens in humans and sets a low upper limit to the human breast cancer risk produced by the conditions of exposure that have operated in the past. Individual components of oral contraceptives are broadly carcinogenic in the laboratory, but the laboratory evidence relating to the specific combinations present in oral contraceptives is not sufficient to override the epidemiological evidence. It is believed that epidemiological surveillance of oral contraceptives should be continued.

# HAIR DYES

# HAIR DYES:

# LABORATORY EVIDENCE

### W. G. FLAMM

*Office of Toxicological Sciences,
Bureau of Foods,
Food and Drug Administration,
Washington DC, USA*

## INTRODUCTION

In the mid-1970s, concern over coal-tar hair dyes came to public attention largely because of the studies performed by Dr Bruce Ames and co-workers (Ames *et al.*, 1975), who found that a substantial number of hair-dye preparations contain dye ingredients that were determined to be mutagenic to *Salmonella typhimurium*. These concerns, coupled with knowledge that such substances can penetrate the skin, led to their extensive review and testing.

In the spring of 1977, the IARC (1978) reviewed the then available data on the most commonly used hair-dye ingredients. These included: 4-amino-2-nitrophenol, 2,4-diaminoanisole, 1,2-diamino-4-nitrobenzene, 1,4-diamino-2-nitrobenzene, 2,4-diaminotoluene, 2,5-diaminotoluene, *meta*-phenylenediamine and *para*-phenylenediamine. The IARC Working Group found that, with few exceptions, none of these compounds had been adequately tested, and, therefore, they could not be evaluated for carcinogenicity. Since that time, many hair-dye ingredients have been tested at the National Cancer Institute in life-time, or virtually life-time, feeding studies (using both rats and mice) in which the hair-dye ingredient was added to the animals' diet. In most instances, the highest level produced some weight loss or reduction in weight gain or exhibited other toxic effects, but usually without adversely affecting the long-term survival of the animals.

## RESULTS

Of the compounds mentioned above, three gave negative results in tests conducted at the National Cancer Institute: 1,2-diamino-4-nitrobenzene, 2,5-diaminotoluene and *para*-phenylenediamine. 1,2-Diamino-4-nitrobenzene was not, in the judgement of reviewers at the National Cancer Institute, tested at a high enough dosage level. The other two compounds that gave negative results did appear to have been adequately tested and were thus assumed to be noncarcinogenic under the conditions of the test, using male and female Fischer 344 rats and male and female $B6C3F_1$ hybrid mice.

Four of the above cited hair-dye ingredients were found to be clearly carcinogenic, inducing a variety of neoplasms, depending upon the sex, the species and the specific aromatic amine compound administered. 4-Amino-2-nitrophenol (National Cancer Institute, 1978a) was clearly carcinogenic for male Fischer 344 rats, producing transitional-cell carcinomas of the urinary bladder in a dose-related manner. The $p$ value for the dose-related trend test was $<0.001$. It also appeared to be active in female Fischer rats; however, in this case, the carcinomas were insufficient in number to give a statistically significant result. No evidence was found for its carcinogenicity in mice. It was, as are most aromatic nitro compounds, strongly mutagenic in the *Salmonella* test. The levels fed in the diet were approximately 0.1–0.25%, and it is interesting to compare these levels with those used in hair-dyeing formulations, which range from 0.1–1% in shampoo-type preparations for dyeing hair in which 4-amino-2-nitrophenol is used. Usually, the material is left on the hair or scalp for only 20–40 minutes before being removed by rinsing. Since 4-amino-2-nitrophenol is a non-permanent or semi-permanent hair dye, it is used with relative frequency—perhaps as often as once or twice a week.

The other nitro compound found to be carcinogenic is 2-nitro-*para*-phenyl-enediamine, which is used in both permanent and semi-permanent hair dyes. The importance of this latter distinction is that the permanent hair dyes are used with oxidizing agents which cause them to react within the hair shaft. These hair dyes are ordinarily used infrequently—no more than once or twice a month. As a consequence, all else being constant, less human exposure would be expected from permanent hair dyes than from semi-permanent ones. 2-Nitro-*para*-phenylenediamine is strongly mutagenic in *Salmonella* strain TA1535 and in *Escherichia coli* WP2 (National Cancer Institute, 1979a) and is measurably mutagenic in L5178Y at the thymidine kinase locus (Palmer *et al.*, 1976). It also induced neoplastic transformation in 10T½ cells, which are derived from C3H mice.

In female B6C3 hybrid mice, a dose-related increase in the incidence of hepatocellular neoplasms was observed. These were primarily diagnosed as adenomas. No statistically or biologically significant increase in the incidence of tumours at any other site was observed in these mice, nor was any increase in incidence found among male B6C3 hybrid mice. Negative results were also obtained in both male and female rats of the Fischer 344 strain. On the basis of the induction of hepatocellular neoplasms, reviewers at the National Cancer Institute concluded that, under the conditions of the test, 2-nitro-*para*-phenylenediamine is carcinogenic to female B6C3 mice. The dietary levels used were from 0.05–0.1% in male rats, 0.1–0.2% in female rats and 0.2–0.4% in mice. How these exposure levels compare to human exposure from hair dyeing is, of course, extremely difficult to estimate, except that actual systemic exposure of mice and rats to 2-nitro-*para*-phenyl-enediamine in these experiments must have been thousands of times greater than that which derives from hair dyeing.

2,4-Diaminotoluene is used only as a permanent hair dye. It was found in a National Cancer Institute study (National Cancer Institute, 1978b) to induce hepatocellular carcinomas and adenomas in male and female Fischer rats and in female B6C3F$_1$ mice. Carcinomas and adenomas were induced in the mammary gland of both female rats and mice. Also, in female mice, there was a suggestion that the incidence of lymphomas may have been increased by treatment. The levels administered in the diet were approximately 0.01–0.02%. While these levels are relatively low in comparison to those of the other hair-dye compounds reported to be carcinogenic, they are, in all probability, many thousands of times higher, in terms of systemic exposure, than that expected to occur from hair dyeing.

2,4-Diaminoanisole, also a permanent hair-dye constituent, was shown in National Cancer Institute studies (National Cancer Institute, 1979b) to be carcinogenic in both rats and mice. In male and female rats, it induced neoplasms of the skin and associated glands. In male and female mice, it induced follicular-cell thyroid tumours, and in male and female rats both follicular- and C-cell tumours of the thyroid. The levels fed were 0.1–0.5% for rats and 0.1–2% in mice. Again, it is not possible to estimate the exposure of humans to 2,4-diaminoanisole from hair-dye use; nevertheless, the US Food and Drug Administration published a final rule on 16 October 1979 (Food and Drug Administration, 1979) requiring a warning on all cosmetic products containing this hair dye.

## CONCLUSIONS

Four different hair dyes have collectively produced a variety of neoplasms of the bladder, liver, thyroid, lymphatic system and skin. All four of these aromatic amine dyes are mutagenic, particularly the two nitro aromatic dyes, which are transformed metabolically to ultimate carcinogens by the various metabolic systems used with bacterial tester strains.

## REFERENCES

Ames, B.N., Kammen, H.O. & Yamaski, E. (1975) Hair dyes are mutagenic: identification of a variety of mutagenic ingredients. *Proc. natl Acad. Sci. USA, 72,* 2423–2427

Food and Drug Administration (1979) Cosmetic product warning statements; coal tar hair dyes containing 2,4-diaminoanisole. *Fed. Reg., 44,* 59509–59512

IARC (1978) *IARC Monographs on the Evaluation of the Carcinogenic Risk of Chemicals to Man,* Vol. 16, *Some Aromatic Amines and Related Nitro Compounds —Hair Dyes, Colouring Agents and Miscellaneous Industrial Chemicals,* Lyon, pp. 43–142

National Cancer Institute (1978a) Bioassay of 4-amino-2-nitrophenol for possible carcinogenicity. *NCI tech. Rep. Ser. No. 94,* pp. 1–35

National Cancer Institute (1978b) Bioassay of 2,4-diaminoanisole sulfate for possible carcinogenicity. *NCI tech. Rep. Ser. No. 84,* pp. 1–64

National Cancer Institute (1979a) Bioassay of 2-nitro-p-phenylenediamine for possible carcinogenicity. *NCI tech. Rep. Ser. No. 169,* pp. 1–44

National Cancer Institute (1979b) Bioassay of 2,4-diaminotoluene for possible carcinogenicity. *NCI tech. Rep. Ser. No. 162,* pp. 1–41

Palmer, K.A., DeNunzio, A. & Green, S. (1976) The mutagenic assay of some hair dye components using the thymidine kinase locus of L5178Y mouse lymphoma cells. *Toxicol. appl. Pharmacol.,* **37,** 108

# HAIR DYES:

# EPIDEMIOLOGICAL EVIDENCE

### L. KINLEN

*CRC Cancer Epidemiology Unit, Edinburgh, Scotland, UK*

## INTRODUCTION

Among the first reported results of the Ames test (Ames *et al.*, 1975), none was more disturbing than the finding that many permanent and semi-permanent hair dyes are strongly mutagenic in this test system. These preparations are in widespread use (over 40% of women in the USA are estimated to have used some type of hair-care preparation), and a hazard to man had not previously been seriously considered. Moreover, systemic absorption is known to occur from their use (Kiese & Rascher, 1968).

The suggestion that certain hair dyes might be carcinogenic in man has led to a number of studies, and the main purpose of this paper is to review these briefly. They belong to two broad types:
  (i) studies of cancer mortality and incidence among hairdressers and beauticians—occupational groups with particular exposure to hair dyes—and
  (ii) studies of hair-dye use and cancer among individuals, mainly by the case-control approach.

## OCCUPATIONAL STUDIES

Occupational studies of hairdressers form an indirect approach to the problem presented by hair dyes. Only studies that concern female hairdressers are considered here, since their exposure to hair dyes is so much greater than that among their male counterparts. The main findings of these studies are summarized in Table 1. The most extensive data available on cancer mortality in this group come from the routine occupational mortality surveys carried out by the Registrar General for England and Wales in the periods around the decennial censuses. Most of the relevant analyses refer to single women, since married women are usually classified according to the

Table 1. Studies of cancer in female hairdressers and beauticians[a]: observed (O) to expected (E) ratios

| Period covered | Region | Deaths or cases | Total cancers | | Breast cancer | | Lung cancer | | Cervical cancer | | Reference |
|---|---|---|---|---|---|---|---|---|---|---|---|
| | | | O/E | (O) | O/E | (O) | O/E | (O) | O/E | (O) | |
| 1949–1953 | England/Wales | D[b] | 1.27 | (47) | 1.44 | (13) | 2.00 | (4) | 4.00 | (4) | Registrar General (1958) |
| 1959–1963 | England/Wales | D[b] | 1.08 | (65) | 1.75 | (21) | 1.00 | (3) | 1.50 | (3) | Registrar General (1971) |
| 1970–1972 | England/Wales | D[b] | 1.03 | (34) | 1.14 | (8) | 1.00 | (2) | 3.00 | (3) | Office of Population Censuses and Surveys (1978) |
| 1958–1962 | California, USA | D | 1.09 | (24) | – | (5) | 6.00 | (6) | – | (4) | Garfinkel et al. (1977) |
| 1972–1975 | California, USA | C | 1.00 | (135) | 1.00 | (43) | 1.82 | (20) | – | – | Menck et al. (1977) |
| 1953–1977 | Japan | D | 1.06 | (148) | 0.62 | (5) | 1.29 | (9) | – | – | Kono et al. (1983) |

[a] Not included is a Danish study (Clemmesen, 1977), see page 60
[b] Restricted to single women
–, Not given

occupation of their husbands. These analyses involve comparisons of the numbers of death certificates involving a mention of the occupational category in question, with the numbers expected if the mortality rates among all single women had applied to the single female hairdressers recorded as currently employed in the relevant census. Differences in occupational definitions at census and at death registration, particularly with respect to whether the current or a previous occupation is stated, obviously affect the validity of the findings. Nevertheless, it is noteworthy that only in the earliest period considered (1949–1953), when exposure to hair dyes was presumably less than it is now, was there any appreciable excess of deaths from cancer (47 observed, 37 expected). In the two more recent periods (1959–1963 and 1970–1972), the number of deaths from cancer was close to that expected. Of particular types of cancer for which data have been published, the most striking excess concerns breast cancer in the period 1959–1963, when 21 deaths were recorded compared to only 12 expected. In the latest examination, however, for the years 1970–1972, no excess was observed (8 observed, 7 expected). The only cancer recorded as showing a consistent excess in these data is cervical cancer, and this does not persist after adjusting for social class.

An examination of 3460 death certificates of adult females in Alameda County, California, USA, for the years 1958–1962 revealed 24 beauticians. This was contrasted with an expected number of 14, estimated on the basis of an examination of a smaller number (1000) of non-cancer death certificates. Of specific sites of cancer, the only significant excess was lung cancer, with six deaths recorded against about one expected (Garfinkel *et al.*, 1977). Another study in California also found an excess of lung cancer among female beauticians but of smaller magnitude, with 20 cases registered in the Los Angeles County Tumor Registry compared to 11 expected if the total of 135 cancers in such women had been distributed similarly to cancers in all women with stated occupations in the registry. As the authors stressed, no information was available about the smoking habits of beauticians in California (Menck *et al.*, 1977).

The most detailed study of cancer among hairdressers is a follow-up study of all but 0.4% of 7736 female Japanese beauticians from 1953 until the end of 1977 (Kono *et al.*, 1983). No significant excess of deaths from cancer was recorded (148 observed, 139 expected); and, among specific sites of cancer, the only significant excess was from stomach cancer, with 61 deaths observed against 46 expected, unadjusted for social class. There was a deficiency of deaths from breast cancer and no significant excess from lung cancer.

Two other follow-up studies of female beauticians have been carried out, both in the USA, but their results have not been presented in detail in the literature. In the first, identifying details of cohorts of beauticians and teachers in Connecticut were matched against the records of the Connecticut Cancer Registry (Walrath, 1977). The only cancer to show an appreciable difference between these occupational groups (of similar size and age structure) was acute leukaemia, with seven cases observed among beauticians compared to no case recorded among teachers. However, when expected numbers of cases of acute leukaemia were calculated on the basis of women–years at risk among the beauticians (with no contribution by untraced women beyond their last date of registration in the State), there was no longer a significant excess. It is also relevant that in the three occupational mortality analyses of single hairdressers in England and Wales referred to in Table 1, there was a consistent deficiency of deaths from leukaemia (0, 3 and 1 observed compared to 2, 7 and 3 expected, respectively). The other prospective study that covered beauticians is a large American Cancer Society study, in which no excess of cancer at any site was found in this occupational group (Hammond, 1977).

In addition to the above, an examination has been reported of cancers among hairdressers registered in the Danish Cancer Registry, in which females showed an excess for each site for the calendar period 1943–1972. The consistency of these excesses strongly suggests that they reflect systematic differences in occupational definitions used at census (from which expected figures were derived) and at cancer registration (the observed figures), the former perhaps stressing current or full-time and the latter embracing former or part-time occupations (Clemmesen, 1977).

## STUDIES OF HAIR-DYE USE BY INDIVIDUALS

Most of the studies that have directly studied hair-dye use by individuals with cancer have been of the case-control type, but, in addition, there was a large cross-sectional survey among 120 557 US nurses. These studies have mainly concerned permanent and semi-permanent dyes, although other hair-care preparations, such as rinses, do contain aromatic agents that are carcinogenic to animals (IARC, 1978). The cancers that have received most attention have been cancers of the breast and bladder.

*Breast cancer*

Breast cancer has received particular attention, largely because it is the most frequent serious malignancy in women in many countries, but also because an early report from New York state claimed that a much greater proportion of affected women had used hair dyes than a comparison group (Shafer & Shafer, 1976). In fact, evaluation of that study was not possible from the limited details provided. No less than 87 of 100 patients with breast cancer were said to have been 'long-time users of hair colouring', whereas only 26% of an unstated number of unaffected women of similar age were 'regular users of permanent hair dyes'. No detail was given about the selection of cases or controls, and details of hair-dye use by women with breast cancer were sometimes obtained from relatives.

Details of other studies of breast cancer and permanent and semi-permanent hair dyes are shown in Table 2. None of the case-control studies found any significant relationship between hair-dye use and breast cancer overall (Kinlen *et al.*, 1977; Shore *et al.*, 1979; Nasca *et al;* 1980; Stavraky *et al.*, 1981). However, in one study (Shore *et al.*, 1979), a relationship was found with a measure of cumulative hair-dye use (number of years used times frequency per year). It is relevant, however, that this was significant only after a multivariate score was used to control for confounding, a procedure that is known to produce misleading results on occasions (Pike *et al.*, 1979). It may also be relevant that there were differences in the collection of details about hair dyes between the cases and controls, since in no less than 23% of the cases but

Table 2. Studies of breast cancer and permanent and semi-permanent hair dyes

| Reference | Type[a] | Controls | Cases[b] | | Controls[b] | | Relative risk |
|---|---|---|---|---|---|---|---|
| | | | No. | % HD users | No. | % HD users | |
| *Case-control studies* | | | | | | | |
| Kinlen et al. (1977) | Int | Hospital | 191 | 38 (14) | 561 | 38 (15) | 0.95 NS |
| Shore et al. (1979) | Tel | Screening clinic | 129 | 33 (18) | 193 | 37 (16) | 1.08 NS |
| Nasca et al. (1980) | Tel | Population | 118 | 49[c] | 233 | 44 | 1.28 NS |
| Stavraky et al. (1981) | Int | Hospital and neighbourhood | 85 | 56 | 170 | 53 | 1.30 NS |
| *Cross-sectional survey* | | | | | | | |
| Hennekens et al. (1979) | Mail | Nurses | 1 131 | 24 (8) | 117 019 | 32 | 1.06 NS |

[a] Int, interview; Tel, telephone
[b] HD, hair-dye; figures in parentheses give the percentage of women exposed to hair dyes 10 or more years before diagnosis
[c] Includes rinses
NS, not significant

in only 8% of the controls the respondent to the telephone interview was a relative and not the woman herself. Moreover, the breast cancers in this study had been diagnosed up to 15 years prior to these interviews (mean, 7.5 years).

Two other case-control studies involving 118 cases (Nasca et al., 1980) and 85 cases (Stavraky et al., 1979), respectively, indicated no overall relationship between hair-dye use and breast cancer. The first of these studies, however, found a significant relationship in the small subgroup of cases and controls who had reported a previous history of benign breast disease (30 cases, 16 controls, among whom 19 cases and 5 controls had used hair dyes). This difference implied a relative risk of 4.5 (confidence limits, 1.2–16.8). It is relevant, however, that no prior hypothesis linked hair dyes to benign breast disease, and no support for this relationship was found in a smaller study (Stavraky et al., 1981).

By far the largest study of hair dyes and cancer is that reported by Hennekens and his colleagues (1979), who received details of hair-dye use and previous cancers from 120 557 married female US nurses. Among these, 3538 cases of cancer were recorded, and the numbers of affected women who had used hair dyes before the diagnosis of cancer were compared with expected numbers derived internally on the assumption that there was no relationship between the reporting of cancer and previous hair-dye use. No excess of breast cancer among hair-dye users was found (270 observed, 258 expected). Among women who had used permanent dyes for more than 20 years before the onset of cancer, there was an excess of breast cancer, with 24 observed compared to 16 expected. However, among those who had first used these dyes 16 to 20 years before diagnosis, there was a similar deficit in numbers of this cancer (16 observed, 25 expected).

*Bladder cancer*

Four case-control studies of hair-dye use and bladder cancer have been reported (Jain et al., 1977; Neutel et al., 1978; Stavraky et al., 1981; Hartge et al., 1982), and details are summarized in Table 3. No relationship was found between hair-dye use and bladder cancer in any of these studies. They include a large study by Hartge and her colleagues (1982), in which 2982 cases of bladder cancer and 5782 controls were interviewed in different parts of the USA—perhaps the largest case-control study of a specific cancer ever conducted in one country. In this study, no relationship was found between hair-dye use and bladder cancer, even among those individuals who had begun dyeing their hair 20 or more years before diagnosis of cancer.

*Other cancers*

Small numbers of cancers at certain sites other than the breast and bladder were included in a case-control study in Canada, namely endometrium (36 cases), cervix (58), lung (70) and lymphomas and leukaemias (70). No significant relationship was found between hair-dye use and any of these sites of cancer. Larger numbers of these and other cancers were identified by Hennekens and his colleagues (1979) in their large prevalence survey among US nurses (Table 4). A significant excess was recorded for two sites—the cervix uteri and the vagina and vulva. A relationship between hair-

Table 3. Studies of bladder cancer and permanent and semi-permanent hair dyes

| Reference | Sex | Cases | | Controls | | Relative risk | Confidence limits |
|---|---|---|---|---|---|---|---|
| | | No. | HD[a] users (%) | No. | HD[a] users (%) | | |
| Case-control studies | | | | | | | |
| Jain et al. (1977) | M + F | 107 | — | 107 | — | 1.1 | (0.4–3.0) |
| Neutel et al. (1978) | M + F | 50 | 36 | 50 | 38 | 1.0 | — |
| Stavraky et al. (1981) | F | 85[b] | 56 | 170 | 53 | 1.1 | (0.4–2.8) |
| Hartge et al. (1982) | M | 2 249 | 8 | 4 270 | 7 | 1.1 | (0.9–1.4) |
| Hartge et al. (1982) | F | 733 | 61 | 1 498 | 58 | 0.9 | (0.8–1.1) |
| Cross-sectional survey | | | | | | | |
| Hennekens et al. (1979) | F | 37[b] | 14 | 117 019 | 32 | 0.6 | — |

[a] HD, hair-dye
[b] Includes other urinary organs

Table 4. Observed and expected numbers of cancers by site among women who had used permanent hair dyes[a]

| Site of cancer | Observed (O) | Expected (E) | Risk ratio[b] |
|---|---|---|---|
| Breast | 270 | 258.2 | 1.06 |
| Skin—non-melanoma | 166 | 156.9 | 1.07 |
| —melanoma | 23 | 23.7 | 0.96 |
| Ovary and tubes | 20 | 10.9 | 1.02 |
| Corpus uteri | 64 | 59.4 | 1.10 |
| Cervix uteri | 124 | 92.4 | 1.44** |
| Other lower genital tract | 10 | 5.2 | 2.58* |
| Thyroid | 26 | 25.8 | 1.01 |
| Lymphoma | 10 | 15.7 | 0.59 |
| Upper gastrointestinal tract | 3 | 3.1 | 0.96 |
| Lower gastrointestinal tract | 15 | 17.8 | 0.80 |
| Connective tissue | 9 | 6.2 | 1.56 |
| Urinary tract | 5 | 7.4 | 0.62 |
| Upper respiratory tract | 8 | 6.4 | 1.32 |
| Lower respiratory tract | 5 | 4.8 | 1.06 |
| Other | 15 | 14.5 | 0.98 |
| Total | 773 | 716.4 | 1.10* |

[a] From Hennekens et al. (1979)

[b] $\frac{\text{O/E among users}}{\text{O/E among non-users}}$

\* $p = 0.02$
\*\* $p < 0.001$

dye use and cigarette smoking was noted in this study as in a previous one (Kinlen et al., 1977). When the calculations were adjusted for the effects of cigarette smoking, the magnitude of these positive associations with genital cancers was reduced, but they remained statistically significant. The risk of cervical cancer did not increase appreciably with increasing intervals since hair dyes were first used.

## CONCLUSION

Hairdressers and beauticians must have greater exposure to hair dyes than other occupational groups, and it is also likely that they dye their own hair more than the average. It is hardly surprising that they should have attracted study, since, even though the extent of individual exposure is usually unknown, concern about the possible carcinogenicity of hair dyes arose from experimental work that did not point to any specific site of cancer in man. In contrast to case-control studies, such cohort studies allow the investigation of cancer of all sites—an important advantage when there is no clue as to the specific site of cancer that may be implicated. It is clear that any increased risk of cancer that might be found among hairdressers is not necessarily due to exposure to hair dyes, for they are also exposed to other possible carcinogens, such as aerosol propellants. But, in fact, no specific cancer has consistently been found to be increased in incidence among hairdressers and beauticians. The excess of lung cancer noted in two small Californian studies might simply reflect the smoking habits of these women, since a relationship between hair-dye use and smoking has been noted in both the UK and the USA. In British occupational mortality analyses, an approximately two-fold excess of cervical cancer among single women in these groups is apparent only if the standard of comparison is all single women and does not persist after adjustment for social class.

Studies of hair-dye use itself have also given largely negative results. Cancers of the bladder and breast have been the subject of most studies, the former with wholly negative, the latter with largely negative results. Exceptions in the case of breast cancer have been limited and consist in one study of a positive relationship that depended on a method of controlling for confounding of uncertain validity. In another study, a positive finding was restricted to a subgroup of women with previous benign breast disease—a group not linked with hair-dye use by any prior hypothesis. For other cancers, the only positive finding has concerned cancers of the cervix, vagina and vulva in a prevalence survey involving large numbers of US nurses. Whereas the excess of cervical cancer might reflect other aspects of their life style, the excess of vaginal and vulval cancers is more intriguing, since little is known of their etiology, and further studies would be of value.

While recognizing that the numbers of individuals studied with exposure to hair dyes for over 20 years is relatively limited, the existing data that relate to cancer are largely reassuring.

## REFERENCES

Ames, B.N., Kammen, H.O. & Yamasaki, E. (1975) Hair dyes are mutagenic: identification of a variety of mutagenic ingredients. *Proc. natl Acad. Sci. USA, 72,* 2423–2427

Clemmesen, J. (1977) Statistical studies in the aetiology of malignant neoplasms. V. Trends and risks. Denmark 1943–72. *Acta pathol. microbiol. scand. (Suppl.), 261,* 1–286

Garfinkel, J., Selvin, S. & Brown, S.M. (1977) Brief communication: possible increased risk of lung cancer among beauticians. *J. natl Cancer Inst., 58,* 141–143

Hammond, E.C. (1977) *Some negative findings (polio, small pox, tetanus and diphtheria vaccines, beauticians) and evaluation of risks.* In: *American Cancer Society's 19th Science Writers' Seminar, Sarasota, Florida,* New York, American Cancer Society

Hartge, P., Hoover, R., Altman, R., Austin, D.F., Cantor, K.P., Child, M.A., Key, C.R., Mason, T.J., Marrett, L.D., Nyers, M.H., Narayana, A.S., Silverman, D.T., Sullivan, J.W., Swanson, G.M., Thomas, D.B. & West, D.W. (1982) Use of hair dyes and risk of bladder cancer. *Cancer Res., 42,* 4784–4787

Hennekens, C.H., Speizer, F.E., Rosner, B., Bain, C.J., Belanger, C. & Peto, R. (1979) Use of permanent hair dyes and cancer among registered nurses. *Lancet, i,* 1390–1393

IARC (1978) *IARC Monographs on the Evaluation of the Carcinogenic Risk of Chemicals to Man*, Vol. 16, *Some Aromatic Amines and Related Nitro Compounds—Hair Dyes, Colouring Agents and Miscellaneous Industrial Chemicals*, Lyon

Jain, M., Morgan, R.W. & Elinson, L. (1977) Hair dyes and bladder cancer. *Can. med. Assoc. J., 117*, 1131–1132

Kiese, M. & Rascher, E. (1968) The absorption of *p*-toluenediamine through human skin in hair dyeing. *Toxicol. appl. Pharmacol., 13*, 325–331

Kinlen, L.J., Harris, R., Garrod, A. & Rodriguez, K. (1977) Use of hair dyes by patients with breast cancer: a case-control study. *Br. med. J., ii*, 366–368

Kono, S., Tokudome, S., Ikeda, M., Yoshimura, T. & Kuratsune, M. (1983) Cancer and other causes of death among female beauticians. *J. natl Cancer Inst., 70*, 443–446

Menck, H.R., Pike, M.C., Henderson, B.E. & Jing, J.S. (1977) Lung cancer risk among beauticians and other female workers: brief communication. *J. natl Cancer Inst., 59*, 1423–1425

Nasca, P.C., Lawrence, C.E., Greenwald, P., Chorost, S., Arbuckle, J.T. & Paulson, A. (1980) Relationship of hair dye use, benign breast disease, and breast cancer. *J. natl Cancer Inst., 64*, 23–28

Neutel, C.I., Nair, R.C. & Last, J.M. (1978) Are hair dyes associated with bladder cancer? *Can. med. Assoc. J., 119*, 307–308

Office of Population Censuses and Surveys (1978) *The Registrar General's Decennial Supplement, England and Wales 1970–1972, Occupational Mortality*, London, Her Majesty's Stationery Office, p. 124

Pike, M.C., Anderson, J. & Day, N. (1979) Some insights into Miettinen's multivariate confounder score approach to case-control study analyses. *J. Epidemiol. Community Health, 33*, 104–106

Registrar General (1958) *The Registrar General's Decennial Supplement, England and Wales, 1951, Occupational Mortality*, Part 2, Vol. 2, *Tables*, London, Her Majesty's Stationery Office, pp. 154–291

Registrar General (1971) *The Registrar General's Decennial Supplement, England and Wales, 1961, Occupational Mortality Tables*, London, Her Majesty's Stationery Office, p. 185

Shafer, N. & Shafer, R.W. (1976) Potential of carcinogenic effects of hair dyes. *N.Y. State J. Med., 76*, 394–396

Shore, R.E., Pasternack, B.S., Thiessen, E.V., Sadow, M., Forbes, R. & Albert, R.E. (1979) A case-control study of hair dye use and breast cancer. *J. natl Cancer Inst., 62*, 277–283

Stavraky, K.M., Clarke, E.A. & Donner, A. (1979) Case-control study of hair dye use by patients with breast cancer and endometrial cancer. *J. natl Cancer Inst., 63*, 941–945

Stavraky, K.M., Clarke, E.A. & Donner, A. (1981) A case-control study of hair-dye use and cancers of various sites. *Br. J. Cancer, 43*, 236–239

Walrath, J. (1977) *Cancer Incidence among Cosmetologists*, PhD dissertation, New Haven, CT, Yale University

# HAIR DYES:

## CONCLUSION

### Rapporteur: P. FRASER

Initially, the discussion focused on the findings in a large cross-sectional survey of US nurses. The unusual distribution of tumour types was noted, suggesting selection bias in the tumours actually reported, one explanation being that only survivors were able to complete the questionnaire. A common etiology was suggested as a possible explanation for the significant excesses of cancers of both the cervix uteri and of the vagina and vulva, but the data are difficult to interpret in the absence of information on other factors known to increase the risk of these cancers.

A case-control study of melanoma was described by Dr Armstrong (Holman & Armstrong, 1983), which revealed a fairly clear dose-response relationship, in terms of recorded number of applications of semi-permanent and temporary hair dyes, and Hutchinson's melanotic freckle (a sub-type of melanoma), but no association with use of permanent hair dyes. The association with semi-permanent hair dyes was not explained by the known association with exposure to sunlight. This finding is interesting in view of the fact that the semi-permanent hair dyes in shampoos tend to contain nitro-aromatic amines and are applied more often than permanent hair dyes.

Hair dyes have been widely used since the 1920s, but information on first use and frequency and duration of use is scanty. They are a heterogenic group of compounds, which have become over the years increasingly more complicated and have been reformulated to minimize the level of potentially carcinogenic ingredients. Thus, any attempt at subdivision beyond the distinction between permanent and semi-permanent hair dyes, for example, is unlikely to be profitable.

Some hair-dye constituents have induced a variety of cancers in animals in life-time feeding experiments. The relevance of these observations to human exposure is uncertain, given that hair dyes are used topically and intermittently and, even if absorbed, will result in systemic exposures many thousands of times lower than those required to produce tumours in animals. Topical application of hair dyes has not produced tumours at the site of application in animals.

No specific cancer has been found consistently to be increased in incidence in groups occupationally exposed to hair dyes, such as hairdressers and beauticians. Studies of hair-dye use have also given largely negative results, but an apparent excess of cervical, vaginal and vulval cancers in US nurses warrants more detailed investigation in case-control studies. Further studies of the association between hair-dye use and Hutchinson's melanotic freckle would also be of value.

While the results of these epidemiological studies are largely reassuring, it should be recognized that the exposure data on which they are based are very limited. In particular, because the composition of hair dyes has been so variable, the findings cannot be related to exposure to any particular brand or specific constituent(s) of hair dyes. Furthermore, the studies provide no information about the safety of newer agents introduced in recent years.

The evidence regarding the carcinogenicity of hair dyes is inadequate to permit a firm conclusion but suggests that, as a group, these agents are unlikely to have produced a quantitatively large increase in risk under the conditions of exposure that have operated in the past.

## REFERENCE

Holman, C.D.J. & Armstrong, B.K. (1983) Hutchinson's melanotic freckle melanoma associated with non-permanent hair dyes. *Br. J. Cancer*, **48**, 599–601

# HYDRAZINE

# HYDRAZINE:

# LABORATORY EVIDENCE

### J.R.P. CABRAL

*Unit of Mechanisms of Carcinogenesis, Division of Experimental Carcinogenesis, International Agency for Research on Cancer, Lyon, France*

Hydrazine was first discovered in 1887, but it did not become a significant commercial chemical until the Second World War, when it was used as a rocket propellant. It has a very wide range of uses: in the preparation of fuels, agricultural chemicals (e.g., maleic hydrazide) and blowing agents and in the manufacture of medicinals (e.g., isoniazid, nitrofurazone, phenylhydrazine).

The evidence for the activity of hydrazine in short-term tests was reviewed recently (Table 1; IARC, 1982). It was shown to be mutagenic and gave indirect evidence of DNA repair in bacteria and fungi in the absence of an exogenous metabolic activation system. It did not induce mutation in cultured mammalian cells, either in the presence or absence of metabolic activation; conflicting results have been obtained with regard to the induction of chromosomal aberrations, sister chromatid exchanges and DNA repair. In two separate studies, hydrazine caused cell transformation (in the baby hamster kidney assay). In several experiments in mice, hydrazine did not induce bone-marrow micronuclei or morphological abnormalities in sperm; in one study it did not increase the level of sister chromatid exchanges in bone marrow. No data on humans were available. According to IARC criteria, the evidence for activity in these tests was considered to be sufficient.

The evidence for the carcinogenicity of hydrazine in experimental animals has also been examined by the IARC (1974, 1982). It was shown to be carcinogenic (Table 2) to mice following its oral administration, producing lung and liver tumours, and after its intraperitoneal administration, producing lung tumours, sarcomas and myeloid leukaemias. The carcinogenicity of hydrazine was also shown in rats: following oral administration, it produced lung and liver tumours. In more recent studies (Table 3), rats of both sexes and male hamsters exposed daily by inhalation to 5 ppm hydrazine developed nasal tumours. After repeated exposure by inhalation to a dose of 1 ppm hydrazine, rats developed nasal turbinate tumours and male mice developed pulmonary adenomas. The incidence of nasal turbinate tumours in rats was dose-

Table 1.  Hydrazine: evidence for activity in short-term tests[a]

|  | DNA damage | Mutation | Chromosomal anomalies | Other[b] |
|---|---|---|---|---|
| Prokaryotes | + | + |  |  |
| Fungi/Green plants | + | + | + |  |
| Insects |  |  |  |  |
| Mammalian cells (in vitro) | ? | − | ? | T(+) |
| Mammals (in vivo) |  |  | − |  |
| Humans (in vivo) |  |  |  |  |

[a] From IARC (1982)
[b] T, cell transformation

Table 2. Experimental evidence of carcinogenicity of hydrazine administered by the oral route[a]

| Species | Range of daily doses (mg/kg) | Results |
|---|---|---|
| Mouse | 2–44 | Lung and liver tumours<br>Does not initiate skin carcinogenesis<br>Reduction of mammary tumour incidence in C3H mice |
| Rat | 30–45 | Lung and liver tumours |
| Hamster | 23–30 | No carcinogenicity |

[a] From IARC (1974, 1982)

Table 3. Experimental evidence of carcinogenicity of hydrazine administered by inhalation[a]

| Species | Dose | Results |
|---|---|---|
| Mouse (female) | 1 ppm (1.3 mg/m$^3$) | Lung adenomas |
| Rat | 1–5 ppm (1.3–6.5 mg/m$^3$) | Nasal tumours (dose-related) |
| Hamster (male) | 5 ppm (6.5 mg/m$^3$) | Nasal tumours |

[a] From MacEwen et al. (1980) (abstract)

related. The increased tumour incidences in mice and hamsters occurred only with maximum tolerated dose levels (MacEwen et al., 1980).

According to IARC criteria, there is sufficient evidence for the carcinogenicity of hydrazine in experimental animals (IARC, 1982). As mentioned above, this compound is used in the manufacture of the pesticide maleic hydrazide (MH), and

Table 4. Comparison of two studies on infant mice with the pesticide maleic hydrazide[a]

| Reference | Strain | Hydrazine impurity (ppm) | Length of experiment | Sex | Total TBA[b] | | Mice with LCT[c] | |
|---|---|---|---|---|---|---|---|---|
| | | | | | Treated | Controls | Treated | Controls |
| Cabral & Ponomarkov (1982) | C57BL/6 | 1 | Lifespan | F | 24/36 | 13/22 | 0/36 | 0/22 |
| | | | | M | 26/41 | 14/23 | 7/41 | 3/23 |
| Epstein & Mantel (1982) | Swiss | 4 000 | 50 weeks | F | 0/43 | 0/68 | 0/43 | 0/68 |
| | | | | M | 19/26 | 4/48 | 19/26 | 4/48 |

[a] Maleic hydrazide dissolved in tricaprylin was given subcutaneously on days 1, 7, 14 and 21 of life at doses of 5, 10, 20 and 20 mg/mouse, respectively.
[b] TBA, tumour-bearing animals
[c] LCT, liver-cell tumours

an important problem that I would like to draw to your attention is the presence of varying amounts of hydrazine as an impurity in many products (e.g., MH) available on the market. Recently, we looked into the potential carcinogenicity of a sample of MH that contained less than 1 ppm hydrazine as an impurity (Table 4): after subcutaneous administration of this sample to infant mice, no significant increase in the incidence of liver-cell tumours was noted (Cabral & Ponomarkov, 1982). In an earlier test with MH containing 4000 ppm hydrazine as an impurity, however, a marked increase in the incidence of liver-cell tumours was noted (Epstein & Mantel, 1968). We believe that the difference in the hydrazine content of the two samples of MH was an important factor in the results obtained.

## REFERENCES

Cabral, J.R.P. & Ponomarkov, V. (1982) Carcinogenicity study of the pesticide maleic hydrazide in mice. *Toxicology*, **24**, 169–173

Epstein, S.S. & Mantel, N. (1968) Hepatocarcinogenicity of the herbicide maleic hydrazide following parenteral administration to infant Swiss mice. *Int. J. Cancer*, **3**, 325–335

IARC (1974) *IARC Monographs on the Evaluation of Carcinogenic Risk of Chemicals to Man*, Vol. 4, *Some Aromatic Amines, Hydrazine and Related Substances, N-Nitroso Compounds and Miscellaneous Alkylating Agents*, Lyon, pp. 127–136

IARC (1982) *IARC Monographs on the Evaluation of the Carcinogenic Risk of Chemicals to Humans*, Suppl. 4, *Chemicals, Industrial Processes and Industries Associated with Cancer in Humans. IARC Monographs, Volumes 1 to 29*, Lyon, pp. 136–138

MacEwen, J.D., Vernot, E.H. & Haun, C.C. (1980) Chronic inhalation toxicity of hydrazine: oncogenic effects (Abstract 297). *Proc. Am. Assoc. Cancer Res.*, **21**, 74

# HYDRAZINE:

# EPIDEMIOLOGICAL EVIDENCE

### N.J. WALD[1]

*Imperial Cancer Research Fund Cancer Epidemiology and Clinical Trials Unit, Radcliffe Infirmary, Oxford, UK*

## INTRODUCTION

Hydrazine ($N_2H_4$) is a colourless, fuming, oily liquid with an ammonia-like odour. The estimated total world production in 1981 was 30 000 tonnes. About 75% of hydrazine is used as a chemical intermediate in the production of pesticides or plastic additives; about 10% is used in fine chemical manufacture (particularly in the production of isoniazid and allopurinol); and the remaining proportion is used as a deoxygenating material for boiler-feed water and as a propellant in rocketry.

Hydrazine has been categorized by the IARC (1982) as being probably carcinogenic to humans and by the American Conference of Governmental Industrial Hygienists (1980, 1981) as an industrial substance suspected of carcinogenic potential for man. The evidence is based on the administration of hydrazine and its sulfate salts to animals (Biancifiori *et al.*, 1966; Juhász *et al.*, 1966, 1967; Severi & Biancifiori, 1968; Biancifiori, 1970; Toth, 1972; Bhide *et al.*, 1976; MacEwen *et al.*, 1981). A separate paper deals with the evidence from studies in experimental animals.

There is one published epidemiological study of exposure to hydrazine and cancer. It was carried out to see whether there was any evidence that hydrazine was carcinogenic to man and to see what had happened to men who had been exposed to hydrazine vapour in the course of their work. The study was conducted by the ICRF Unit in Oxford in collaboration with a British chemical manufacturer, and the first report was published recently (Wald *et al.*, 1984). This paper is based closely on that report.

---

[1] Present address: Department of Environmental and Preventive Medicine, St Bartholomew's Hospital Medical College, University of London, UK

## DESCRIPTION OF THE PLANT

At a factory in the East Midlands between 1945 and 1971, about 700 tonnes of hydrazine were produced per year. The plant was in an enclosed building and, as hydrazine was not considered to be more hazardous than ammonia, exhaust ventilation was not provided. Hydrazine was kept in open tanks, and hydrazine compounds were made from strong hydrazine solutions in open vessels, which were heated to enhance evaporation. Packing of hydrazine into small commercial containers was carried out in the same building. Spillages were not necessarily flushed away immediately, and it was the practice of the laboratory workers to pipette hydrazine solutions by mouth.

## LEVEL OF EXPOSURE

Unfortunately, no measurement of atmospheric hydrazine was ever carried out at this plant, although the level of hydrazine is likely to have been 1–10 ppm in the general plant area, and levels much higher than this (up to 100 ppm) may have occurred close to hydrazine storage vessels. These estimations have been derived by simulation of spillages and from calculations using data on the saturated vapour pressure of hydrazine at 20 °C, which would suggest that maximum levels of 100 ppm are possible.

A number of other organic chemicals were manufactured in the same factory. Production of hydrazine at the plant ended in 1971, and the factory was closed in 1973.

## STUDY POPULATION, CATEGORIES OF EXPOSURE AND FOLLOW UP

Between 1945 and 1971, 427 men were known to have been employed at the plant for at least six months. For each of these men, the following information was sought: identifying details, date of birth, date of first employment, date of leaving the company, and an estimate of the extent of hydrazine exposure, based on the knowledge of the factory works manager.

Each type of employment was classified into one of the following categories, according to the estimated degree of exposure:

Category 1 – exposure associated with the direct manufacture of hydrazine, or of its derivatives, or involving the use of liquid hydrazine as a raw material. Exposure to hydrazine vapour was likely to have been greatest for men in this category, who may have been exposed to about 1–10 ppm in the ambient air.

Category 2 – exposure associated with an incidental presence in an area of the plant concerned with the manufacture of hydrazine or its derivatives (e.g., fitters, engineers). Exposure in this category was unlikely to have been to more than 1 ppm, and was probably to less than 0.5 or 1 ppm for most of their employment.

Category 3 – little or no exposure. Men in this category were unlikely to have been exposed to hydrazine more than slightly and then only infrequently.

The men were followed to the end of July 1982 through the cooperation of the Office of Population Censuses and Surveys by flagging their National Health Service records in the National Health Service Central Register at Southport.

Table 1. Numbers of men exposed and man-years at risk by category of exposure, duration of exposure and years since first exposure

| Category | Duration of exposure (months) | Years since first exposure | No. of men | No. of man-years |
|---|---|---|---|---|
| 1 | 6–23 | < 10 | 78 | 350 |
|   |      | ≥ 10 | 73 | 198 |
|   | 24 or more | < 10 | 54 | 508 |
|   |            | ≥ 10 | 50 | 509 |
|   | 6 or more | All | 78 | 1565 |
| 2 or 3 | 6 or more | All | 375 | 6786 |
| All | All | All | 427[a] | 8351 |

[a] Men who were first exposed in Categories 2 or 3 and who were subsequently exposed in Category 1 contributed man-years at risk in Categories 2 or 3 initially and to Category 1 after their first exposure in that category. The numbers of men in each category, therefore, add up to more than 427 in all, as some men contributed to more than one category. Similarly, all men who contributed man-years at risk more than 10 years after first exposure and for durations of exposure of two years or more, also contributed to man-years at risk less than 10 years after first exposure and to less than two years duration of exposure.

It was possible to trace 406 (95%) of the 427 men. The 21 untraced men were excluded from the study from the latest date they were known to have been living at their last known address, or, in the case of four men for whom this date was missing, from the last date of their employment in the factory. Men who changed from one category of exposure to another were considered to have been at risk in relation to the highest category to which they had previously been exposed (see footnote to Table 1).

## RESULTS

Table 1 shows the number of man-years under observation according to the category of exposure, duration of exposure, and the number of years since first exposure. Exposure in Category 1 accounted for 19% of the 8351 man-years at risk.

Tables 2–5 show the numbers of deaths observed compared with the numbers that would have been expected if the men had experienced the same death rates as those of men of the same ages in the same years in England and Wales as a whole, subdivided similarly (five-year age groups and five-year calendar periods). The observed mortality is close to that expected for lung cancer (Table 2), other cancers (Table 3) and all other causes (Table 4), irrespective of the category of exposure. No death occurred from nasal cancer, a tumour caused in rats by exposure to hydrazine vapour (MacEwen et al., 1981). Two men with the heaviest exposure died of lung cancer, compared with 1.61 expected. Both had first been exposed more than 10 years previous to when they developed the disease; one had been exposed for two months, the other for 16 years.

Table 2. Numbers of deaths observed and expected according to category of exposure, duration of exposure and years since first exposure. Mortality due to lung cancer

| Category of exposure | Duration of exposure (months) | Years since first exposure | No. of deaths Observed | No. of deaths Expected |
|---|---|---|---|---|
| 1 | 6–23 | < 10 | 0 | 0.05 |
|   |      | ≥ 10 | 1 | 0.10 |
|   | 24 or more | < 10 | 0 | 0.31 |
|   |            | ≥ 10 | 1 | 1.15 |
|   | 6 or more | All | 2 | 1.61 |
| 2 or 3 | 6–23 | All | 3 | 5.03 |
| All | 6 or more | All | 5 | 6.65 |

Table 3. Numbers of deaths observed and expected according to category of exposure, duration of exposure and years since first exposure. Mortality due to cancers other than lung cancer

| Category of exposure | Duration of exposure (months) | Years since first exposure | No. of deaths Observed | No. of deaths Expected |
|---|---|---|---|---|
| 1 | 6–23 | < 10 | 0 | 0.12 |
|   |      | ≥ 10 | 0 | 1.16 |
|   | 24 or more | < 10 | 0 | 0.42 |
|   |            | ≥ 10 | 0 | 1.48 |
|   | 6 or more | All | 0 | 2.18 |
| 2 or 3 | 6 or more | All | 7 | 7.10 |
| All | 6 or more | All | 7 | 9.27 |

## DISCUSSION

The number of men exposed to hydrazine in this study was small. Only 78 men had had substantial exposure, estimated at between 1–10 ppm hydrazine vapour in air, and observations were made on only 707 man-years at risk for 10 years or more (up to a maximum of 36 years) after first exposure. Table 6 shows the relative risk (observed/expected) and corresponding 95% confidence intervals according to category of exposure and cause of death. From the point of view of the purpose of this Symposium, it is the upper bound on the confidence interval that is relevant. The results are in general encouraging, in that none of the upper-bound estimates are greater than 2, but a note of caution is introduced by the upper-bound estimate of 4.5 relating to men who had first been exposed in the heaviest category more than 10 years previously.

Table 4. Numbers of deaths observed and expected according to category of exposure, duration of exposure and years since first exposure. Mortality due to causes other than cancer

| Category of exposure | Duration of exposure (months) | Years since first exposure | No. of deaths Observed | Expected |
|---|---|---|---|---|
| 1 | 6–23 | < 10 | 0 | 0.60 |
|   |       | ≥ 10 | 0 | 0.70 |
|   | 24 or more | < 10 | 1 | 2.00 |
|   |       | ≥ 10 | 7 | 7.39 |
|   | 6 or more | All | 8 | 10.69 |
| 2 or 3 | 6 or more | All | 29 | 34.86 |
| All | 6 or more | All | 37 | 45.55 |

Table 5. Numbers of deaths observed and expected according to category of exposure, duration of exposure and years since first exposure. Mortality due to all causes

| Category of exposure | Duration of exposure (months) | Years since first exposure | No. of deaths Observed | Expected |
|---|---|---|---|---|
| 1 | 6–23 | < 10 | 0 | 0.77 |
|   |       | ≥ 10 | 1 | 0.96 |
|   | 24 or more | < 10 | 1 | 2.73 |
|   |       | ≥ 10 | 8 | 10.02 |
|   | 6 or more | All | 10 | 14.48 |
| 2 or 3 | 6 or more | All | 39 | 46.99 |
| All | 6 or more | All | 49 | 61.47 |

# REFERENCES

American Conference of Governmental Industrial Hygienists (1980) *Documentation of the Threshold Limit Values,* Cincinnati, OH

American Conference of Governmental Industrial Hygienists (1981) *Threshold Limit Values for Chemical Substances and Physical Agents in the Workroom Environment, with Intended Changes, for 1981,* Cincinnati, OH

Bhide, S.V., D'Souza, R.A., Sawai, M.M. & Ranadive, K.J. (1976) Lung tumour incidence in mice treated with hydrazine sulphate. *Int. J. Cancer, 18,* 530–535

Table 6. Relative risk (observed deaths/expected deaths) of specified cause of death according to category of exposure

| Cause of death | Category of exposure | Relative risk | 95% confidence interval of relative risk |
|---|---|---|---|
| Lung cancer | 1 | 1.2 | 0.2–4.5 |
| | 2 or 3 | 0.6 | 0.1–1.7 |
| Other cancer | 1 | 0.0 | 0.0–1.7 |
| | 2 or 3 | 1.0 | 0.4–2.0 |
| Other causes | 1 | 0.7 | 0.3–1.5 |
| | 2 or 3 | 0.8 | 0.6–1.2 |
| All causes | 1 | 0.7 | 0.3–1.3 |
| | 2 or 3 | 0.8 | 0.6–1.1 |

Biancifiori, C. (1970) Hepatomas in CBA/Cb/Se mice and liver lesions in golden hamsters induced by hydrazine sulfate. *J. natl Cancer Inst.*, **44**, 943–949

Biancifiori, C., Giornelli-Santilli, F.E., Milia, V. & Severi, L. (1966) Pulmonary tumours in rats induced by oral hydrazine sulphate. *Nature*, **212**, 414–415

IARC (1982) *IARC Monographs on the Evaluation of the Carcinogenic Risk of Chemicals to Humans*, Suppl. 4, *Chemicals, Industrial Processes and Industries Associated with Cancer in Humans (IARC Monographs, Volumes 1 to 29)*, Lyon, pp. 136–138

Juhász, J., Baló, J. & Szende, B. (1966) Tumour-inducing effect of hydrazine in mice. *Nature*, **210**, 1377

Juhász, J., Baló, J. & Szende, B. (1967) Carcinogenic properties of hydrazine. *Magy. Onkol.*, **11**, 31–36

MacEwen, J.D., Vernot, E.H., Haun, C.C. & Kinkead, E.R. (1981) *Chronic Inhalation Toxicity of Hydrazine: Oncogenic Effects*, Wright-Patterson Air Force Base, OH, Air Force Aerospace Medical Research Laboratory

Severi, L. & Biancifiori, C. (1968) Hepatic carcinogenesis in CBA/Cb/Se mice and Cb/Se rats by isonicotinic acid hydrazine and hydrazine sulfate. *J. natl Cancer Inst.*, **41**, 331–340

Toth, B. (1972) Tumorigenesis studies with 1,2-dimethylhydrazide dihydrochloride, hydrazine sulfate, and isonicotinic acid in golden hamsters. *Cancer Res.*, **32**, 804–807

Wald, N., Boreham, J., Doll, R. & Bonsall, J. (1984) Occupational exposure to hydrazine and subsequent risk of cancer. *Br. J. ind. Med.*, **41**, 31–34

# HYDRAZINE:

# CONCLUSION

### Rapporteur: L. KINLEN

There was general agreement about the problems of interpreting findings such as those on lung cancer and hydrazine when such small numbers of cases were observed. Moreover, since virtually the only group of British factory workers exposed to hydrazine vapour has already been studied, there is little prospect of extending the study in the UK. Dr Krewski suggested that the negative findings might be due to a combination of the 'healthy worker' effect and also the small number of heavily exposed workers who had been followed up over a long period. Dr Wald agreed, mentioning that only 78 men had had substantial exposure (1–10 ppm hydrazine in air).

There was interest in the smoking habits of the workers reported by Dr Wald, but he pointed out that although many were known to be smokers, data on individuals were limited. No specific rulings about smoking existed in the factory.

In the course of further discussion about the extent of exposure, Dr Wald mentioned that although the particular company in which the work force had been studied recently was still making hydrazine, the conditions of manufacture had changed markedly. Dr Cabral added that in the USA the Environmental Protection Agency, in collaboration with representatives of the relevant industries, held out good prospects for making large reductions in hydrazine levels in factories.

Given the limited prospects for extending the study of workers exposed to hydrazine within the UK, several members pointed out that in this situation an international study would be valuable, since by pooling data more reliable conclusions could be obtained.

The evidence regarding the carcinogenicity of hydrazine is quantitatively too weak to warrant any useful conclusion.

# FORMALDEHYDE

# FORMALDEHYDE:

# LABORATORY EVIDENCE

W.G. FLAMM & V. FRANKOS

*Office of Toxicological Sciences,
Bureau of Foods,
Food and Drug Administration,
Washington DC, USA*

## INTRODUCTION

Formaldehyde is used in the production of specific formaldehyde resins and in a wide variety of other industries and occupations. Occupational exposures to formaldehyde have been common and, in some instances, occurred at relatively high levels. Auerbach (1951) postulated that formaldehyde would be mutagenic under the proper circumstances and demonstrated that casein treated with formaldehyde was strongly mutagenic to *Drosophila melanogaster*. These factors have no doubt contributed to the fact that, over the past three decades, numerous studies have been conducted in animals to determine whether formaldehyde is carcinogenic (see Table 1). Subcutaneous injection studies in rats were the first to suggest that formaldehyde might be a carcinogen. Several ingestion studies, however, gave negative results (Watanabe *et al.*, 1954; Watanabe & Sugimoto, 1955; Brendel, 1964; Della Porta *et al.*, 1968, 1970). Direct application of formalin to the palate of rabbits gave evidence that formalin could induce cancer (Mueller *et al.*, 1978); however, an early formaldehyde inhalation study with mice gave negative results (Horton *et al.*, 1963). Hamsters exposed to formaldehyde vapour for life did not show a neoplastic response (Dalbey, 1981). Ultimately, an inhalation study in rats showed that formaldehyde had a significant carcinogenic effect (Pavkov *et al.*, 1981); this finding was later confirmed by Albert *et al.* (1982). The two independent studies in rats showed a significant number of nasal squamous-cell carcinomas in exposed animals.

A working group convened by IARC (1982) concluded that there is sufficient evidence that formaldehyde gas is carcinogenic to rats, but that the available epidemiological studies provided inadequate evidence to assess the carcinogenicity of formaldehyde in man.

Table 1. Studies of the carcinogenicity of formaldehyde in laboratory animals

| Species | No. animals per group | Compound and dose | Route and frequency | Length of study | Tumour incidence |
|---|---|---|---|---|---|
| Rat | 10 | 1 ml of 0.4% formalin | Subcutaneous once/week | 15 months | 2/10 local sarcomas 1/10 liver 1/10 peritoneal cavity |
| Rat | 20 | 1 ml of 9–40% HMT[a] | Subcutaneous once/week | Until tumour formation | 7/20 local sarcomas |
| Rat (BD) | 15/sex | 0, 0.4 g HMT (total dose) | Oral, drinking-water | 333 days | Negative |
| Mouse (CTM, SWR, C3HF) | 27–100/sex | 0, 0.5, 1.0, 5.0% HMT | Oral, drinking-water | 110–130 weeks | Negative |
| Rat (Wistar) | 12–48/sex | 0, 1.0, 5.0% HMT | Oral, drinking-water | 156 weeks | Negative |
| Rat (Wistar) | 12–24/sex | 0, 1.0% HMT | Oral, drinking-water for 20–40 weeks | 3 generations | Negative |
| Rat (Wistar) | 16–48/sex | 0, 2.0% HMT | Oral, drinking-water for 50 weeks | 2 years | Negative |
| Rabbit | 6 | 0, 3% formalin | Direct application to palate 5 times/week | 10 months | 1/6 'carcinoma in situ' of the palate |
| Mouse (C3H) | 42–60/group | 0, 41, 80, 163 ppm formaldehyde | Inhalation 1 h/day 3 days/week | 35–70 weeks | Negative |
| Hamster (Syrian golden) | 88–132 males | 0, 10 ppm formaldehyde | Inhalation 5 h/day 3 days/week | Lifetime | Negative |
| Rat (Fischer 344) | 120/sex | 0, 2.0, 5.6, 14.3 ppm formaldehyde | Inhalation 6 h/day 5 days/week | 2.5 years | Nasal tumours (combined sexes, 103/200 at 14.3 ppm, 2/214 at 5.6 ppm) |
| Mouse (B6C3F$_1$) | 120/sex | 0, 2.0, 5.6, 14.3 ppm formaldehyde | Inhalation 6 h/day 5 days/week | 2.5 years | Nasal tumours (2/240 at 14.3 ppm) |
| Rat (Sprague-Dawley) | 100 males | 0, 14.7 ppm formaldehyde + 10.6 ppm HCl | Inhalation 6 h/day 5 days/week | Lifetime | Nasal tumours (25/99) |
| Rat (Sprague-Dawley) | 100 males | 0, 14.3 ppm formaldehyde + 10.0 ppm HCl | Inhalation 6 h/day 5 days/week | 588 days[b] | Nasal tumours (12/100) |
| Rat (Sprague-Dawley) | 100 males | 0, 14.1 ppm formaldehyde + 9.5 ppm HCl | Inhalation 6 h/day 5 days/week | 588 days[b] | Nasal tumours (6/100) |
| Rat (Sprague-Dawley) | 100 males | 0, 14.2 ppm formaldehyde | Inhalation 6 h/day 5 days/week | 588 days[b] | Nasal tumours (10/100) |

[a] HMT, Hexamethylenetetramine; forms formaldehyde upon decomposition
[b] Interim results

## RESULTS

A recently released report by the Chemical Industry Institute of Toxicology (Pavkov *et al.*, 1981) gives the results of a study of inhalation exposure to formaldehyde for 24 months in rats and mice. Both species were exposed to formaldehyde concentrations of 0, 2.0, 5.6, and 14.3 ppm for six hours a day on five days a week. Each exposure group consisted of 120 male and 120 female animals of each species. Scheduled interim sacrifices were done at 6, 12, 18 and 24 months, during the exposure period. Mortality unrelated to formaldehyde exposure was substantial among the male mice as a consequence of fighting. Mortality was also increased in rats exposed to 14.3 ppm formaldehyde.

Mice in all exposure groups did not exhibit a statistically significant compound-related increase in the incidences of neoplasms, except that two male mice exposed to 14.3 ppm had squamous-cell carcinomas of the nasal cavity. In addition, there were increased incidences of epithelial dysplasia and squamous-cell metaplasia in male mice exposed to 5.6 and 14.3 ppm and in female mice exposed to 14.3 ppm.

A highly significant increased incidence of nasal cavity squamous-cell carcinomas occurred in both male (51 of 101) and female (52 of 99) rats exposed to 14.3 ppm formaldehyde. One male and one female rat exposed to 5.6 ppm also had squamous-cell carcinomas of the nasal cavity. In all groups of exposed rats there were increased incidences of epithelial dysplasia and squamous-cell metaplasia of the nasal cavity. Females exposed to 14.3 ppm also had increased incidences of squamous metaplasia of the trachea. Rats exposed to 14.3 ppm had increased incidences of squamous epithelial hyperplasia and squamous atypia. There was an apparent regression of the nasal cavity squamous-cell metaplasia three months after exposure ended in the 2.0- and 5.6-ppm exposure groups.

Discussing the lesions found in the nasal cavity of exposed rats the authors stated that: 'Apparent progression from squamous metaplasia to squamous epithelial hyperplasia with increased keratin production and then to areas of squamous papillary hyperplasia with areas of cellular atypia was evident in the high dose group only. More advanced lesions diagnosed as carcinoma "in situ" and, finally, invasive squamous cell carcinomas of the nasal turbinates were present in rats from the intermediate and high exposure groups, but were statistically different from controls in the high exposure group only.'

Two experiments reported by Albert *et al.* (1982) confirm the above finding that formaldehyde inhalation can induce nasal cancers in rats. Of the two lifetime studies reported, one has been completed and the second has progressed through 588 days. The first study was performed to evaluate the effects of chronic inhalation exposure to a mixture of formaldehyde and hydrogen chloride (HCl), because these compounds can combine in gaseous form to form bischloro(methyl)ether (BCME), a known animal and human carcinogen. The second study examined exposure to formaldehyde and HCl together and separately to understand how each compound contributes to the effects caused by combining the compounds. Both studies involved male Sprague-Dawley rats exposed for six hours a day on five days a week, until natural death or sacrifice when moribund.

In the first study, HCl and formaldehyde gases were mixed before passage into the exposure chamber at a 75-fold dilution. The average measured chamber concentrations of HCl and formaldehyde were 10.6 ppm and 14.7 ppm, respectively, and the concentration of BCME was approximatedly 1.0 ppb. In this study, 100 rats were exposed to the formaldehyde-HCl mixture, 50 served as air sham-exposed controls, and another 50 were untreated.

In the second study, 100 rats were used in each of five exposure groups: (1) premixed formaldehyde and HCl, (2) HCl and formaldehyde added directly to the exposure chamber, (3) HCl only, (4) formaldehyde only and (5) air sham-exposed controls. Nominal chamber concentrations of HCl and formaldehyde were about 10 ppm and 14 ppm, respectively, for each treatment group. The BCME level was not determined.

All nasal cancers appeared as nasal swellings and were confirmed histologically. The first nasal cancer appeared after 223 days in the first study and after 420 days in the second. In the first study, 71 treated and 8 untreated control rats showed epithelial hyperplasia and hyperplasia with atypia of the nasal cavity. Hyperplasia of the larynx, trachea and bronchi occurred at roughly equal incidences in control and treated rats.

The incidence of nasal squamous-cell carcinoma in the first study was 25%; no similar tumour was found in the controls. Although BCME (which has been shown to produce nasal tumours) was detected in the exposure chambers in this study, it is unlikely that it was involved in the formation of the squamous-cell carcinomas, since it normally induces neurogenic carcinomas (mainly aesthesioneuroepitheliomas), which were not found in this study. The second study also indicates that the probable carcinogen was formaldehyde and not BCME. The numbers of squamous-cell carcinomas in the two groups exposed to the combination of formaldehyde and HCl (where BCME would be present) were 12 and 6, respectively, which were similar to the number of tumours found in the group exposed to formaldehyde only (10). At the time the interim report was presented, none of the animals that had been exposed to HCl alone was reported to have a nasal tumour.

These studies indicate that formaldehyde is carcinogenic to the nasal epithelium of rats. Mice and hamsters may be less sensitive to the carcinogenic effect, since they appear to be less sensitive to the direct cytotoxic (irritative) effect of formaldehyde. It is evident that formaldehyde is carcinogenic to rats and probably to mice at the site of initial contact and does not cause tumours at distant sites. Of the species tested, the rat is the only species in which a statistically significant number of tumours arose.

The formaldehyde-related neoplasms were accompanied by other tissue lesions. Sequential progression of these lesions (hyperplasia, metaplasia and dysplasia) to neoplasia may indicate that they are necessary precursors of the neoplasia. If this is true, the carcinogenesis may occur through epigenetic mechanisms. It has been argued (Weisburger & Williams, 1980) that such a mechanism may include a threshold dose level for expression of carcinogenesis. Thus, the nasal epithelium of the hamster, which appears to be less sensitive to the cytotoxic effects of formaldehyde than the nasal epithelium of the rat (Ben-Dyke et al., 1980), would be less likely to develop squamous-cell carcinomas.

Spangler and Ward (1983) have tested formaldehyde for its ability to function as a promoter, using the Sencar mouse back skin as a test model. 7,12-Dimethylbenz[a]anthracene was deployed as the 'initiator' carcinogen, and, although the study was still incomplete when reported, the authors indicated there was only a slight possibility that formaldehyde 'may be a very weak promoting agent'. Another interesting possibility regarding the carcinogenicity of formaldehyde is raised by a recent study (Ohshima et al., 1984) in which an easily nitrosatable amine, thiazolidine 4-carboxylic acid, is reported to be formed by reaction of formaldehyde with cysteine in vivo and in vitro. Given the simplicity of the formaldehyde molecule, it should be possible eventually to determine with a high degree of accuracy the mechanism by which it produces cancer in rats. Such information should provide 'clues' and 'markers' for accessing, on both qualitative and quantitative grounds, the degree to which exposure of man to formaldehyde represents a risk of cancer.

## CONCLUSIONS

Formaldehyde has been tested over the past 30 years in several mammalian species for carcinogenicity using many different routes of exposure. Until recently, no convincing evidence of its carcinogenicity had been reported. Two recent, independent inhalation studies have demonstrated conclusively that relatively high concentrations of formaldehyde in air produce invasive squamous-cell carcinomas of the nasal turbinates in exposed rats. A parallel study in mice also resulted in the induction of carcinomas of the nasal turbinates. While formaldehyde is mutagenic, the mechanism of its carcinogenic action is unclear and is currently the subject of much speculation.

## REFERENCES

Albert, R., Sellakumar, A., Laskin, S., Kuschner, M., Nelson, N. & Snyder, C. (1982) Gaseous formaldehyde and hydrogen chloride induction of nasal cancer in the rat. *J. natl Cancer Inst.*, **68**, 597–603

Auerbach, C. (1951) Some recent results with chemical mutagens. *Hereditas*, **37**, 1–16

Ben-Dyke, R., Rusch, G. & Hogan, G. (1980) *A 26-Week Inhalation Toxicity Study of Formaldehyde in the Monkey, Rat, and Hamster (Project No. 79-7259. Final Report)*, East Millstone, NJ, Bio-Dynamics, Inc.

Brendel, R. (1964) Hexamethylenetetramine tolerance in rats (Ger.). *Arzneimittel. Forsch.*, **14**, 51–53

Dalbey, W.E. (1981) *Effects of Formaldehyde or Nitrogen Dioxide on Tumors in Hamster Respiratory Tract. Toxicology in Review (OSHA Formaldehyde Docket H-225)*

Della Porta, G., Colnaghi, M.I. & Parmiani, G. (1968) Non-carcinogenicity of hexamethylenetetramine in mice and rats. *Food Cosmet. Toxicol.*, **6**, 707–715

Della Porta, G., Cabral, J.R. & Parmiani, G. (1970) Transplacental toxicity and carcinogenesis studies in rats with hexamethylenetetramine (Ital.). *Tumori*, **56**, 325–334

Horton, A.W., Type, R. & Stemmer, K.L. (1963) Experimental carcinogenesis of the lung. Inhalation of gaseous formaldehyde or an aerosol of coal tar by C3H mice. *J. natl Cancer Inst.*, **30**, 31–43

IARC (1982) *IARC Monographs on the Evaluation of the Carcinogenic Risk of Chemicals to Humans*, Vol. 29, *Some Industrial Chemicals and Dyestuffs*, Lyon, pp. 345–389

Mueller, R., Raabe, G. & Schumann, D. (1978) Leukoplakia induced by repeated deposition of formalin in rabbit oral mucosa: Long-term experiments with a new 'oral tank'. *Exp. Pathol.*, **16**, 36–42

Ohshima, H., Friesen, M., O'Neill, I. & Bartsch, H. (1983) Presence in human urine of a new *N*-nitroso compound, *N*-nitrosothiazolidine 4-carboxylic acid. *Cancer Lett.*, **20**, 183–190

Pavkov, K.L., Mitchell, R.I., Donofrio, D.J., Kerns, W.D., Connell, M.M., Harroff, H.H., Fisher, G.L., Joiner, R.L. & Thake, D.C. (1981) *Final Report on Chronic Inhalation Toxicology Study in Rats and Mice Exposed to Formaldehyde. Conducted by Battelle Columbus Laboratory, Columbus, Ohio, for Chemical Industry Institute of Toxicology* (Submitted September 18, 1981, revised December 31, 1981)

Spangler, F. & Ward, J.M. (1983) Skin initiation/promotion study with formaldehyde in Sencar mice (in press)

Watanabe, F. & Sugimoto, S. (1955) Studies on the carcinogenicity of aldehydes: Part II. Seven cases of transplantable sarcomas of rats developed in the area of repeated subcutaneous injections of urotropin (hexamethylenetetramine). *Gann*, **46**, 365–367

Watanabe, F., Matsunaga, T., Soejima, T. & Iwata, Y. (1954) Study on aldehyde carcinogenicity: Communication I. Experimentally induced rat sarcomas by repeated injection of formalin. *Gann*, **45**, 451–452

Weisburger, J.H. & Williams, G.M. (1980) *Chemical carcinogens*. In: Doull, J., Klaassen, C.D. & Amdur, M.O., eds, *Casarett and Doull's Toxicology. The Basic Science of Poisons*, 2nd ed., New York, Macmillan, pp. 84–138

# FORMALDEHYDE:

# EPIDEMIOLOGICAL EVIDENCE

### E.D. ACHESON[1]

*MRC Environmental Epidemiology Unit, University of Southampton, Southampton General Hospital, Southampton, UK*

There have been two recent case reports of nasal cancer in persons exposed to formaldehyde: one was an engineer exposed to formaldehyde in the textile industry (Halperin *et al.*, 1983), the other worked in a factory where formaldehyde resins were manufactured (P. Infante & H. Kang, unpublished data). A search of the Danish Cancer Registry for the period 1943–1976 revealed only three cases of nasal cancer among doctors, none of whom were pathologists or anatomists, who are most exposed to formaldehyde (Jensen, 1980). Follow-up studies of British pathologists, over two different time periods, have revealed no excess respiratory cancer of any kind (Harrington & Oakes, 1984). An analysis of cancer mortality by occupation in Ontario Province, Canada, during 1972–1979 recorded no death due to cancer of the nasal cavity among doctors, anatomists and pathologists not medically qualified, morticians and dentists (Kreiger, 1983).

A follow-up of employees of the Monsanto Company in Massachusetts, USA, identified 592 men who had been employed in the factory for a minimum of one year during 1949–1966 and had died during 1950–1976 (Marsh, 1982). Formaldehyde was both produced in the factory and used as a raw material in the manufacture of other substances. Of the 592 men, 136 had had at least one month's employment in areas of the plant where exposure to formaldehyde could have occurred. Proportional mortality ratios were calculated, expected values being derived from proportionate mortality among the US male population as a whole and mortality in Hampden county, where the factory is situated. These latter data, from the county, were not materially different from the national data.

There was no statistically significant excess or deficit of any cause of death among the 136 exposed workers, and overall cancer mortality was lower in these workers than in the remaining 456. There was no death from nasal cancer. The only suggestion

---

[1] Present address: Chief Medical Officer, Department of Health and Social Security, Alexander Fleming House, Elephant & Castle, London, UK

Table 1. Deaths from cancers other than of the respiratory tract in four studies of exposure to formaldehyde

| Study (reference) | Site | No. observed | No. expected | Confidence limits |
|---|---|---|---|---|
| Monsanto chemical plant (Marsh, 1982) | Digestive tract | 2[a]<br>5[a] | 0.5<br>1.6 | 50–1491<br>104– 747 |
| Texas chemical plant (Wong, 1983) | Prostate | 4[a] | 1.2 | 89– 833 |
| New York State morticians (Walrath & Fraumeni, 1983) | Skin<br>Colon<br>Kidney<br>Brain | 8<br>29[a]<br>6[a]<br>6[a] | 3.6<br>20.3<br>2.3<br>2.4 | 96– 438<br>96– 205<br>94– 557<br>90– 533 |
| Du Pont chemical plant (Fayerweather et al., 1983) | Prostate<br>Bladder | Odds ratio<br>Odds ratio | 4.8[a]<br>7.0[a] | 1.0 – 22.4<br>0.61 – 79.9 |

[a] Figures relate to sub-groups, e.g., by age, latent period

of an association with cancer was an excess of cancer of the digestive tract within two sub-groups—men under 45 and whites in whom the latency was more than 20 years and the duration of exposure less than five years. These data are shown in Table 1, together with those from other studies on cancers outside the respiratory tract.

Another industrial follow-up study was based on a chemical plant in Texas, USA, said to be one of the largest producers of formaldehyde in the USA (Wong, 1983). The plant was built in the early 1940s, and the cohort comprised all 2026 white men who had, at some time, been employed there from its beginnings until 1977. These men were also exposed to chemicals other than formaldehyde, such as asbestos, benzene and pigments. The estimated mean duration of exposure in the plant was 11.4 years, and there were 146 deaths. The analysis was based on Standardized Mortality Ratios calculated on the basis of US national rates. The 'healthy worker effect' was reflected in a ratio of 74. There was no statistically significant excess or deficit for any cause of death, and no nasal cancer death. Analysis by latent period showed a small excess mortality from prostatic cancer when the latent period was more than ten years (Table 1).

Following publication of these data, Tabershaw Associates (unpublished data) reported additional data, based on work histories of formaldehyde exposure among the work force. They were able to study a cohort of exposed and unexposed workers, and to conduct a case-control study of exposure history among the dead employees and age-matched controls randomly selected from the remainder of the cohort. No new conclusion came from these additional analyses.

Causes of death among 1132 male morticians in New York State, USA, who were licensed as embalmers during the period 1902–1980 and died during 1925–1980, were analysed (Walrath & Fraumeni, 1983). Formaldehyde is the main preservative in embalming fluids, which contains a variety of other chemicals. Proporationate mortality analysis, with expected values obtained from US national data, showed no statistically significant excess of deaths from all neoplasms combined. There was no nasal cancer death. Subsequent studies of deaths among 1200 morticians in Ontario, Canada, licensed between 1914 and 1967 (Levine et al., 1983), and among 1115 embalmers in California, USA, who died during 1925–1980 (Walrath, 1983), likewise showed no excess of lung cancer and no nasal cancer.

Among the New York State morticians, there were excesses of deaths from skin and colon cancer (Table 1). Four of the eight skin cancer cases were malignant melanoma. The excess of skin cancer increased with both duration of employment in embalming and intensity of exposure (as judged by whether a man was solely an embalmer or directed funerals as well). Among those who only embalmed, there was increased proportional mortality from cancers of the brain and kidney (Table 1).

In 1983, the Du Pont Company (Fayerweather et al., 1983) released the results of a case-control study based on 481 male cancer deaths occurring in eight chemical plants during 1957–1979. The plants either made or used formaldehyde and had done so for at least 15 years. The cases comprised all male cancer deaths occurring among either active or pensioned employees. A limitation of the study was the exclusion of an estimated 15% of the work force who left without a pension or were transferred to a plant not in the study. The controls were an equal number of male employees, pair-matched for age, wage bracket and date of first employment. About one-third of cases had been exposed to formaldehyde, either continuously or intermittently. Almost half of the exposed cases and controls had been exposed for 17 years or more, but they had been exposed to other chemicals as well. Data on smoking history were obtained.

A matched-pair analysis showed no statistically significant elevation in odds ratio of exposure for any cancer site. There was no case of nasal cancer. Lung cancer deaths were examined taking account of smoking history, and no association with formaldehyde exposure was found. This is the only study of formaldehyde in which data on smoking were obtained—albeit the data were crude, being obtained primarily from fellow employees of the deceased. There was an elevation in the odds ratio for two cancer sites, the prostate and bladder (Table 1).

In conclusion, none of the cohort studies or case-control studies has shown an excess of respiratory-tract carcinoma following exposure to formaldehyde, as suggested by the results of animal experiments. In particular, no case of nasal cancer has been reported in any of the follow-up studies. In two case-control studies of nasal cancer—one of 167 patients in Denmark, Finland and Sweden (Hernberg et al., 1983), the other of 160 patients in North Carolina and Virginia, USA (Brinton et al., 1983) —no association was found with occupations in which formaldehyde exposure could have occurred.

Published studies provide no evidence for an association between formaldehyde and lung cancer. Since, unlike the animals in which experimental work has been carried out, man is not an obligatory nose-breather, respiratory exposure to formaldehyde can also take place by the mouth and pharynx. No important excess of cancer at these sites has been found. However, a recent reclassification of deaths in the study at the Monsanto Chemical Plant identified 7 deaths from buccal and pharyngeal cancer as compared with 3.1 expected (P. Infante & H. Kang, unpublished data).

There is some suggestion of associations with carcinoma at other sites. The excess of brain tumour deaths among embalmers in New York State is supported by similar excesses in other studies of professional workers using formaldehyde—namely, embalmers in California, USA (Walrath, 1983), morticians in Ontario, Canada (Levine *et al.*, 1983), pathologists (Harrington & Oakes, 1984; G.M. Matonoski, unpublished data) and anatomists (N. Stroup, unpublished data). The combined results from these studies also suggest an association with leukaemia, and further studies on professional workers using formaldehyde to preserve human tissues are clearly indicated. For cancers at other sites, the evidence of an association with formaldehyde is slight. Three points are raised by Table 1. First, with only one exception (skin cancer in New York State morticians), the excesses of cancer were observed in sub-groups, such as particular age-groups. Second, the numbers of observed cases are small. Third, only two of the results are statistically significant. Since there are some 32 cancer sites, five studies, and analyses by various sub-groups, it is likely that some statistically significant differences would have arisen purely by chance.

The limitations of the studies on formaldehyde are such that they do not provide conclusive evidence regarding the carcinogenicity of formaldehyde to man. These limitations include the small numbers of exposed persons in the studies—especially persons followed up for more than 20 years. For example, in the study in Texas, only 19 cases of cancer were expected in workers among whom the minimum latent period would have been 20 or more years. A further weakness is that the duration of exposure of most persons to formaldehyde has been short. In view of the small numbers, the studies have limited power to detect other than a very large increase in relative risk of a tumour as rare as nasal cancer.

Further studies of populations exposed to formaldehyde are required.

## REFERENCES

Brinton, L., Becker, J., Blot, W., Hoover, R. & Fraumeni, J., Jr. (1983) A case-control study of cancer of the nasal cavity and sinuses (Abstract). *Am. J. Epidemiol.*, **118**, 436

Fayerweather, W.E., Pell, S. & Bender, J.R. (1983) *Case-control study of cancer deaths in DuPont workers with potential exposure to formaldehyde.* In: Clary, J.J., Gibson, J.E. & Waritz, R.S., eds, *Formaldehyde: Toxicology, Epidemiology, and Mechanisms*, New York, Marcel Dekker, pp. 47–113

Halperin, W.E., Goodman, M., Stayner, L., Elliott, L.J., Keenlyside, R.A. & Landrigan, P.J. (1983) Nasal cancer in workers exposed to formaldehyde. *J. Am. med. Assoc.*, **249**, 510–512

Harrington, J.M. & Oakes, D. (1984) Mortality study of British pathologists, 1974–80. *Br. J. ind. Med.*, **41**, 188–191

Hernberg, S., Collan, Y., Degerth, R., Englund, A., Engzell, U., Kuosma, E., Mutanen, P., Nordlinder, H., Sand Hansen, H., Schultz-Larsen, K., Søgaard, H. & Westerholm, P. (1983) Nasal cancer and occupational exposures. Preliminary report of a joint Nordic case-referent study. *Scand. J. Work Environ. Health, 9,* 208–213

Jensen, O.M. (1980) Cancer risk from formaldehyde. *Lancet, ii,* 480–481

Kreiger, N. (1983) Formaldehyde and nasal cancer mortality. *Can. med. Assoc. J., 128,* 248–249

Levine, R.J., Andjelkovich, D.A., Shaw, L.K. & Dallorso, R.D. (1983) *Mortality of Ontario undertakers: a first report.* In: Clary, J.J., Gibson, J.E. & Waritz, R.S., eds, *Formaldehyde: Toxicology, Epidemiology, and Mechanisms,* New York, Marcel Dekker, pp. 127–146

Marsh, G.M. (1982) Proportional mortality patterns among chemical plant workers exposed to formaldehyde. *Br. J. ind. Med., 39,* 313–322

Walrath, J. (1983) Mortality among embalmers (Abstract). *Am. J. Epidemiol., 118,* 432

Walrath, J. & Fraumeni, J.F. (1983) Mortality patterns among embalmers. *Int. J. Cancer, 31,* 407–411

Wong, O. (1983) *An epidemiologic mortality study of a cohort of chemical workers potentially exposed to formaldehyde, with a discussion on SMR and PMR.* In: Gibson, J.E., ed., *Formaldehyde Toxicity,* New York, Hemisphere Publishing Corp., pp. 256–272

FORMALDEHYDE:

# CONCLUSION

Rapporteur: O.M. JENSEN

In considering the reports of the carcinogenic effects of formaldehyde in rats, it seems clear that this animal species is particularly prone to the development of nasal tumours. In comparison with mice, guinea-pigs and rabbits, rats have the relevant sensitivity, anatomy and physiology. It is not at present possible to determine whether the observed species differences are due to variations in sensitivity or can be explained by the existence of a threshold. Although the experiments in rats provide no support for the existence of such a threshold, it was pointed out that the dose-response relationship is cubic rather than linear.

*A priori,* the respiratory organs and the skin are the most likely target organs for formaldehyde carcinogenesis. Formaldehyde is, however, a very reactive gas, and it interacts quickly in a humid atmosphere. The passage of formaldehyde into the lung is facilitated by the presence in the ambient air of particulate matter to which formaldehyde becomes absorbed. If it indeed reaches the lung, there is no reason to doubt that formaldehyde exerts the same carcinogenic action as it does in the proximal airways, if the pertinent concentrations are achieved.

Studies of exposed cohorts in the USA, a case-control study and case series of Danish physicians show no increased risk of lung cancer or nasal cancer associated with exposure to formaldehyde. Some of the follow-up studies have shown increased risks of cancers of the prostate, brain, skin, kidney and bladder. The power of these studies to disclose small increases in relative risk has not been recorded.

In an unpublished retrospective cohort study of workers in six chemical plants in the UK presented by Professor Acheson, no case of nasal cancer has been seen. The statistical power of the study to detect a five-fold increase in risk is 70%, if it is assumed that nasal cancer risk increases immediately upon exposure and if people exposed to all levels are included. If an increased risk occurs only in highly exposed persons 20 years or more after first exposure, then the power to detect a five-fold increase in risk falls to approximately 25%. The study cannot, therefore, be regarded as powerful in ruling out an increase in the risk of nasal cancer.

For all six factories taken together, there is no increased risk of lung cancer when regional variations in lung cancer mortality are taken into account. In one plant, an exposure-response relationship of borderline statistical significance emerges. It must be considered whether it would be more appropriate to examine the trend after pooling the results from all factories studied. The point was also made that the significance of a trend that depends on a low risk in the low-exposure group (standardized mortality ratio, 70) may carry less weight than a trend starting from a standardized mortality ratio of 100, unless there were explanations other than chance for such a low risk among the less heavily exposed.

In the study in the UK, there is no trend in risk associated with duration of exposure among those persons who were heavily exposed; however, the investigators raised the question of whether a sufficient number of persons had been exposed to high levels to see such an effect. It must be noted that short-term workers are those at highest risk; no information exists on occupational background or personal habits. Smoking histories are being investigated, but have not yet been taken into account in the analysis.

This large study lends no support to previous suggestions of associations between formaldehyde exposure and non-respiratory-tract cancers. Four bone tumours have been found, all of which appear to be primary tumours.

In reaching their conclusions about the effects of formaldehyde, the participants of the Symposium considered the published data reviewed by Professor Acheson and the summary of the results of the study carried out by the Medical Research Council of the mortality experience of men occupationally exposed to formaldehyde, which Professor Acheson and his colleagues had presented previously to the management of the individual factories in which the studies had been made and to union representatives. The participants of the Symposium were not able to examine the detailed data, and they would like to have the opportunity to reconsider their views when all the results are available. Meanwhile, they are of the opinion that, while current epidemiological evidence suggests that formaldehyde is unlikely to have produced a quantitatively large increase in the risk of cancer under the conditions of exposure that have operated in the past, current information is inadequate to justify a firm conclusion. As Professor Acheson was still in the course of completing his report on the effects of formaldehyde, he took no part in formulating this assessment.[1]

---

[1] Since this conclusion was formulated, the results of the study by Professor Acheson and his colleagues have been published: Acheson, E.D., Barnes, H.R., Gardner, M.J., Osmond, C., Pannett, B. & Taylor, C.P. (1984) Formaldehyde in the British chemical industry. An occupational cohort study. *Lancet*, *i*, 611–616

**DDT**

# DDT:

# LABORATORY EVIDENCE

## J.R.P. CABRAL

*Unit of Mechanisms of Carcinogenesis, Division of Experimental Carcinogenesis, International Agency for Research on Cancer, Lyon, France*

DDT has been widely used as an insecticide in agriculture and in malaria-control programmes. The insecticidal applications of DDT were identified by Muller in 1939 in Switzerland, and, a few years later, he received the Nobel Prize for his work. By 1943, low-cost production methods had been developed, and commercial production had begun. US production of DDT was greatest in 1963, when 82 million kg were produced (IARC, 1974). At present, there is only one US producer, with a plant capacity of 40 million kg per year. Much of the pesticide manufactured in the US is exported: in 1978, 15 million kg, 51% of which went to India, Thailand and Canada (US Department of Agriculture, 1980).

DDT has been tested for activity in a variety of short-term tests (Table 1). It did not interact with DNA and did not produce unscheduled DNA synthesis in cultured human fibroblasts or in rat, mouse or hamster hepatocytes. DDT was not mutagenic to *Salmonella typhimurium,* to yeast, to cultured rat liver epithelial cells or to human fibroblasts in a rat hepatocyte-mediated assay. It failed to produce chromosomal aberrations in cultured human lymphocytes. It did not produce recessive or dominant lethal mutations in wasps or visible or lethal mutations in mice exposed for five generations. No data on humans were available. The results from these studies were evaluated by an IARC working group in 1982 (IARC, 1982) and were considered to constitute inadequate evidence for activity in short-term tests.

The evaluation of the carcinogenicity of DDT is complicated by the fact that it is stored in tissues, in man and in animals, both as such and in the form of its two metabolites, DDE and TDE. In human fat, the concentration of DDE exceeds that of DDT, at least in the general population. It is therefore best to present separately experimental data on the administration of DDT and those on its metabolite DDE (IARC, 1974; Kreiss *et al.,* 1981).

Data from long-term studies on DDT (Table 2) were reviewed by IARC (1974, 1982). The first suggestion of its carcinogenicity was made by Fitzhugh and Nelson in 1947. After technical DDT was fed to Osborne-Mendel rats for 24 months at

Table 1. DDT: Evidence for activity in short-term tests[a]

|  | DNA damage | Mutation | Chromosomal anomalies | Other[b] |
|---|---|---|---|---|
| Prokaryotes | – | – |  |  |
| Fungi/Green plants |  | – |  |  |
| Insects |  | – |  | DL(–) |
| Mammalian cells (in vitro) | – | – | – |  |
| Mammals (in vivo) |  | – |  | DL(–) |
| Humans (in vivo) |  |  |  |  |

[a] From IARC (1982)
[b] DL, dominant lethal mutations

dietary levels ranging from 100–800 ppm, among the 75 rats still alive after 18 months, 4 had 'low-grade' hepatic-cell carcinomas and 11 had hyperplastic liver nodules. The hepatocarcinogenicity of DDT in rats when given by the oral route has been demonstrated in two other long-term feeding studies (Rossi et al., 1977; Cabral et al., 1982a). No organ other than the liver has been suggested as a target for the carcinogenicity of DDT in rats. The contradictory results obtained in rats may be due to differences in strain susceptibility, sex, dosage levels used, length of treatment or pathological criteria for diagnosis.

In multiple studies in mice, DDT was shown to be carcinogenic following its oral administration, and it almost consistently induced liver tumours in animals of both sexes. The doses that were effective in producing liver tumours ranged from 2 ppm (lowest dietary concentration used) in male CF1 mice (Tomatis et al., 1972) to 250 ppm in male and female BALB/c mice (Terracini et al., 1973). At least two feeding studies also showed an excess of lung adenomas in mice (Shabad et al., 1973; Kashyap et al., 1977); an increase in the incidence of lymphomas was also noted in one of the feeding studies (Kashyap et al., 1977). Following subcutaneous administration to mice, DDT also produced liver tumours and lymphomas (Kashyap et al., 1977).

Four feeding studies at dietary concentrations of up to 1000 ppm in hamsters, carried out in different laboratories, gave consistently negative results (Cabral et al., 1982b; Table 3).

DDE was not mutagenic in the Ames test (Planche et al., 1979; Gold et al., 1981); no other short-term test of DDE has been reported.

In two experiments in mice, DDE was carcinogenic, only to the liver. The dietary doses used in both experiments were in the order of 250 ppm (Tomatis et al., 1974; National Cancer Institute, 1978). In one study in rats with dietary concentrations ranging between 187 and 675 ppm, no carcinogenic effect of DDE was found; however, a marked hepatotoxic effect was noted, together with increased mortality (National Cancer Institute, 1978). In hamsters, in a study conducted at the same time as and with a similar experimental design to one on DDT, which gave negative results, there was an increased incidence of liver tumours (Table 3).

The susceptibility of mice, rats and hamsters to DDT and DDE is compared in Table 4. The qualitative differences in the carcinogenicity of DDT and DDE suggest that the two chemicals should be considered separately when analysing

Table 2. Summary of experiments with DDT given by the oral route to rats

| Reference | Strain | Total no. of treated animals | Max. length treatment (weeks) | Max. dosage (mg/kg per day) | No. survivors in max. dose group at 100 weeks | Evidence of carcinogenicity |
|---|---|---|---|---|---|---|
| Fitzhugh & Nelson (1947) | Osborne-Mendel | 192 | 104 | 40 | 4 | + |
| Treon & Cleveland (1955) | ? | 240 | 104 | 1.2 | 42 | − |
| Kimbrough et al. (1964) | Sherman | 75 | 40 | 2 | ? | − |
| Radomski et al. (1965) | Osborne-Mendel | 60 | 104 | 4 | ? | − |
| Deichmann et al. (1967) | Osborne-Mendel | 60 | 104 | 10 | 40 | − |
| Weisburger & Weisburger (1968) | Fischer 344 | 30 | 52 | 30 | ? | − |
| Rossi et al. (1977) | Wistar | 72 | 152 | 25 | 41 | + |
| National Cancer Institute (1978) | Osborne-Mendel | 200 | 78 | 20–30 | 77 | − |
| Cabral et al. (1982a) | MRC Porton (Wistar) | 196 | 144 | 25 | 27 | + |

Table 3. Summary of carcinogenicity experiments with DDT and DDE in Syrian golden hamsters

| Compound | Number of animals | | Dose (range) | | Duration of treatment (weeks) | Evidence of carcinogenicity | Reference |
|---|---|---|---|---|---|---|---|
| | Experimental | Controls | ppm | mg/kg per day | | | |
| DDT | 115 | 79 | 500–1000 | 40–80 | 44 | None | Agthé et al. (1970) |
| | 180 | 60 | 250–1000 | 20–80 | 78 | None | Graillot et al. (1975) |
| | 200 | 80 | 125–500 | 10–40 | 120 | None | Cabral et al. (1982) |
| | 83 | 91 | 1000 | 80 | >120 | None | Rossi et al. (1980) |
| DDE | 188 | 91 | 50–1000 | 40–80 | >120 | Liver-cell tumours in treated animals of both sexes | Rossi et al. (1980) |

Table 4. Comparative hepatocarcinogenicity of DDT and DDE

|  | Mice | Rats | Hamsters |
| --- | --- | --- | --- |
| DDT | Yes[a] | Yes | No |
| DDE | Yes | No | Yes |

[a] Also lung tumours and lymphomas

correspondences or differences between laboratory animals and humans, also keeping in mind the high accumulation of DDE in adipose tissues of human populations contaminated with DDT.

## REFERENCES

Agthé, C., Garcia, H., Shubik, P., Tomatis, L. & Wenyon, E. (1970) Study of the potential carcinogenicity of DDT in the Syrian golden hamster. *Proc. Soc. exp. Biol. N.Y., 134,* 113–116

Cabral, J.R.P., Hall, R.K., Rossi, L., Bronczyk, S.A. & Shubik, P. (1982a) Effects of long-term intake of DDT on rats. *Tumori, 68,* 11–17

Cabral, J.R.P., Hall, R.K., Rossi, L., Bronczyk, S.A. & Shubik, P. (1982b) Lack of carcinogenicity of DDT in hamsters. *Tumori, 68,* 5–10

Deichmann, W.B., Keplinger, M., Sala, F. & Glass, E. (1967) Synergism among oral carcinogens. IV. The simultaneous feeding of four tumorigens to rats. *Toxicol. appl. Pharmacol., 11,* 88–103

Fitzhugh, O.G. & Nelson, A.A. (1947) Chronic oral toxicity of DDT. *J. Pharmacol. exp. Ther., 89,* 18–30

Gold, B., Leuschen, T., Brunk, G. & Gingell, R. (1981) Metabolism of a DDT metabolite via a chloroepoxide. *Chem.-biol. Interact., 35,* 159–176

Graillot, C., Gak, J.C., Lancret, C. & Truhaut, R. (1975) Studies on the modes and mechanisms of the toxic action of organochlorine insecticides. II. Study on the hamster of the long-term toxic effects of DDT (Fr.). *Eur. J. Toxicol. environ. Hyg., 8,* 353–359

IARC (1974) *IARC Monographs on the Evaluation of Carcinogenic Risk of Chemicals to Man,* Vol. 5, *Some Organochlorine Pesticides,* Lyon, pp. 83–124

IARC (1982) *IARC Monographs on the Evaluation of the Carcinogenic Risk of Chemicals to Humans,* Suppl. 4, *Chemicals, Industrial Processes and Industries Associated with Cancer in Humans (IARC Monographs, Volumes 1 to 29),* Lyon, pp. 105–108

Kashyap, S.K., Nigam, S.K., Karnik, A.B., Gupta, R.C. & Chatterjee, S.K. (1977) Carcinogenicity of DDT in pure inbred Swiss mice. *Int. J. Cancer, 19,* 725–729

Kimbrough, R., Gaines, T.B. & Sherman, J.D. (1964) Nutritional factors, long-term DDT intake and chloroleukemia in rats. *J. natl Cancer Inst., 33,* 215–225

Kreiss, K., Zack, M.M., Kimbrough, R.D., Needham, L.L., Smrek, A.L. & Jones, B.T. (1981) Cross-sectional study of a community with exceptional exposure to DDT. *J. Am. med. Assoc.*, **245**, 1926–1930

National Cancer Institute (1978) Bioassays of DDT, TDE and $p,p'$-DDE for possibile carcinogenicity. *Tech. Rep. Ser. No. 131*

Planche, G., Croisy, A., Malaveille, C., Tomatis, L. & Bartsch, H. (1979) Metabolic and mutagenicity studies on DDT and 15 derivatives. Detection of 1,1-bis($p$-chlorophenyl)-2,2-dichloroethane and 1,1-bis($p$-chlorophenyl)-2,2-trichloroethyl-acetate (kelthane acetate) as mutagens in *Salmonella typhimurium* and of 1,1-bis-($p$-chlorophenyl)ethylene oxide, a likely metabolite, as an alkylating agent. *Chem.-biol. Interact.*, **25**, 157–175

Radomski, J.L., Deichmann, W.B., McDonald, W.E. & Glass E.M. (1965) Synergism among oral carcinogens. I. Results of the simultaneous feeding of four tumorigens to rats. *Toxicol. appl. Pharmacol.*, **7**, 652–656

Rossi, L., Ravera, M., Repetti, G. & Santi, L. (1977) Long-term administration of DDT or phenobarbital-Na in Wistar rats. *Int. J. Cancer*, **19**, 179–185

Rossi, L., Barbieri, O., Cabral, J.R.P., Sanguinetti, M. & Santi, L. (1980) *Carcinogenic bioassay of DDE in hamsters.* In: *Proceedings of the 19th Annual Meeting of the Society of Toxicology,* Washington DC

Shabad, L.M., Kolesnichenko, T.S. & Nikonova, T.V. (1973) Transplacental and combined long-term effect of DDT in five generations of A-strain mice. *Int. J. Cancer*, **11**, 688–693

Terracini, B., Testa, M.C., Cabral, J.R.P. & Day, N. (1973) The effects of long-term feeding of DDT to BALB/c mice. *Int. J. Cancer*, **11**, 747–764

Tomatis, L., Turusov, V., Day, N. & Charles, R.T. (1972) The effect of long-term exposure to DDT on CF-1 mice. *Int. J. Cancer*, **10**, 489–506

Tomatis, L., Turusov, V., Charles, R.T. & Boiocchi, M. (1974) Effect of long-term exposure to 1,1-dichloro-2,2-bis($p$-chlorophenyl)-ethylene, to 1,1-dichloro-2,2-bis-($p$-chlorophenyl)ethane and to the two chemicals combined on CF-1 mice. *J. natl Cancer Inst.*, **52**, 883–891

Treon, J.R. & Cleveland, F.P. (1955) Toxicity of certain chlorinated hydrocarbon insecticides for laboratory animals with special reference to aldrin and dieldrin. *J. Agric. Food Chem.*, **3**, 402–408

US Department of Agriculture (1980) *The Pesticide Review,* Washington DC

Weisburger, J.H. & Weisburger, E.K. (1968) Food additives and chemical carcinogens: on the concept of zero tolerance. *Food Cosmet. Toxicol.*, **6**, 235–242

DDT:

# EPIDEMIOLOGICAL EVIDENCE

J. HIGGINSON

*Universities Associated for Research and Education in Pathology, Inc.,
Bethesda, MD, USA*

## SUMMARY

The use of DDT as a pesticide, until its banning in many countries, is reviewed briefly. The tissue levels of DDT can be regarded as an index of past exposures, and findings in several countries are described. It is emphasized that insufficient case-control studies are available for evaluation and that assessment of the potential carcinogenic effects of DDT in humans is largely dependent on inferences from descriptive epidemiology. DDT is a non-genotoxic carcinogen in animals and a mild hepatoxic agent. The liver is a probable target organ in man. No correlation at the population level can be demonstrated between exposures to DDT and the incidence of cancer at any site. It is concluded that DDT has had no significant impact on human cancer patterns and is unlikely to be an important carcinogen for man at previous exposure levels, within the statistical limitations of the data.

## INTRODUCTION

DDT is an abbreviation for dichlorodiphenoltrichloroethane and represents the prototype of broad-action persistent insecticides. As the first long-acting pesticide, DDT (and its metabolites) has been the subject of intensive study over the last two decades, and it is not easy to add anything new to the existing extensive literature on the subject. Many of the key reports are adequately summarized in a number of publications (Mrak, 1969; IARC, 1974; WHO, 1979). As a widely used pesticide in the community, DDT has aspects of both scientific and historical interest. It is stable under many environmental conditions and resistant to complete breakdown by enzymes in soil microorganisms. Certain of its metabolites, notably DDE, have a stability equal to that of the parent compound, and levels of DDE in tissues are often used as a chemical index of previous exposure to the pesticide.

Following identification of its pesticidal properties in the early 1940s, DDT was tested and then manufactured in vast quantities in the USA. Early production was for use as protection against insect-borne diseases. Following the Second World War, during which control of malaria and typhus among troops and civilians had required comparatively small quantities of DDT, far greater amounts were needed for the control of agricultural and forest pests. As the increasing resistance to DDT of a number of pests became recognized, there was some reduction in use in certain countries, whereas elsewhere there was an increase in use in an attempt to induce control by higher dosage. It is unclear just how much DDT was used (IARC, 1974; WHO, 1979), but, in general, if production in the USA is taken as typical, the quantity produced and used rose massively until about 1960, following which there was a gradual decrease in use in the USA, although production remained high, since much of it was exported. Some restrictions were placed on its use, mainly to minimize residues in food and animal feed. In 1970, the government of Sweden banned the use of DDT, and other governmental agencies elsewhere followed suit shortly thereafter. These actions were justified for a number of reasons, but one of the most cogent was that it represented a threat to human health and could possibly induce cancer, as suggested by the experimental studies outlined by Dr Cabral. Its biological effects are believed to be due to the pesticide itself and to its metabolites, and not to contaminants. In many western countries, including those in North America, the use of DDT has almost ceased. According to the WHO (1979), however, it is still used widely in agriculture and for food protection in many tropical countries. I have not been able to obtain further data on recent usage.

In summary, there was widespread exposure of the general population in many countries to DDT during a period of 20–25 years, following which, in some countries, the level of human exposures declined significantly; however, due to the persistence of the pesticide and its stability in the body, internal exposure has persisted, although at a lower level (WHO, 1979). At the same time, in many countries, a number of subgroups of the population (sprayers, manufacturers, etc.) were exposed to very high levels of DDT. More recently, a number of population groups have been identified as having been exposed to very high levels of the pesticide due to bad waste disposal, e.g. in Triana, Alabama, USA (Kreiss *et al.*, 1981).

## POSSIBLE HEALTH EFFECTS

*Exposure levels*

The possible health effects of DDT can be considered in relation to both unusually highly exposed individuals and the general community. Due to the nature of its usage in agriculture and pest control, DDT was very widely spread and has been identified in almost all regions of the world. For individuals, the major source of DDT in the body has been by ingestion; but, in certain situations, especially occupational, high concentrations may be received through air pollution. It can also apparently be absorbed through the skin. Further information about exposure has been gained from measurements of DDT in body tissues and fluids, which are apparently much better markers of exposure at the cellular or individual level than are ambient environmental

Table 1. Average concentrations of the isomers of DDT and DDE in fat and serum of workers engaged in the manufacture, formulation or use of DDT

| Tissue | No. of men | DDT (mg/kg) | DDE (mg/kg) | Total as DDT (mg/kg) | Estimated exposure (mg/man per day) |
|---|---|---|---|---|---|
| Fat | 1 | 648 | 437 | 1131 | – |
| Urine | 10 | – | – | – | 14 |
| Urine | 16 | – | – | – | 30 |
| Urine | 13 | – | – | – | 42 |
| Fat | 3 | 51 | 44 | 98 | 3.6 |
| Fat | 12 | 74 | 50 | 130 | 6.2 |
| Fat | 20 | 161 | 91 | 263 | 18 |
| Serum | 3 | 0.2113 | 0.1968 | 0.5412 | 6.3 |
| Serum | 12 | 0.1420 | 0.1454 | 0.3584 | 8.4 |
| Serum | 20 | 0.3020 | 0.2719 | 0.7371 | 17.5 |
| Urine | 3 | 0.0165 | 0.0203 | 0.5629 | – |
| Urine | 12 | 0.0145 | 0.0222 | 0.7911 | – |
| Urine | 20 | 0.0145 | 0.0271 | 1.6296 | – |

Table 2. Concentrations of DDT-derived material in body fat of the general population

| Country | Year | No. of samples | Method of analysis[a] | DDT[b] (mg/kg) | DDE as DDT (mg/kg) | Total as DDT[c] (mg/kg) | DDE as DDT (% of total) |
|---|---|---|---|---|---|---|---|
| USA | 1942 | 10 | GLC, colorimetric | ND[d] | ND | ND | |
| USA | 1961–1962 | 28[e] | GLC | 2.4 | 4.3 | 6.7 | 64 |
| USA | 1964–1965 | 42 | GLC | 3.1 | 7.5 | 10.6 | 71 |
| USA | 1966–1968 | 70 | GLC | 1.54 | 5.15 | 6.69 | 77 |
| Israel | 1963–1964 | 254 | Colorimetric | 8.5 | 10.7 | 19.2 | 56 |
| Israel | 1967–1971 | 63 | GLC | 3.0 | 10.6 | 14.4 | 74 |

[a] GLC, gas-liquid chromatography
[b] $p,p'$-DDT and $o,p'$-DDT only
[c] Includes TDE (DDD) and other forms
[d] Not detected
[e] Also tested for DDT and DDE content by a colorimetric method, and the results are included in the samples listed above

measurements. Extensive studies were undertaken during its heyday to identify the amount of the pesticide and of its metabolites in tissues, and data have been accumulated in many countries regarding levels in different organs, fat, sera, etc. DDT is somewhat unusual in that in certain situations levels of human and animal exposure are comparable; therefore its health effects in humans and animals can also be compared at approximately similar levels of exposure.

Some of the results of measurement studies are summarized in Tables 1–7, which are largely drawn from Davies and Edmundson (1972) and WHO (1979). Tables 1 and 2 show the levels in different organs and adipose tissues in workers and in the general population in a number of countries. Of interest is evidence of a socio-economic gradient in the USA and of much higher deposits in blacks than in whites (Davies & Edmundson, 1972; Davies et al., 1972). It has been suggested that this

Table 3. Average DDT contents (ppm total DDT) in fat in males in South Africa and Israel by age[a]

| Male subjects | Age group | | | | | | |
|---|---|---|---|---|---|---|---|
| | Months | Years | | | | | |
| | 0–11 | 1–4 | 5–14 | 15–24 | 25–44 | 45–69 | 70+ |
| South Africa | | | | | | | |
| White | – | 6.89 | – | 5.59 | 14.35 | 6.99 | 7.09 |
| Black | – | 5.26 | 5.69 | 10.35 | 14.11 | 5.45 | – |
| Israel | 6.40 | – | 18.26 | 11.43 | 31.86 | 15.08 | 15.60 |

[a] From IARC (1970)

represents a racial difference in ability to metabolize DDT, as such differences in tissue levels were not found for other pesticides of this type. However, no difference was observed between blacks and whites in South Africa (Tables 3 and 4). Despite cessation of use, DDT levels still remain high in the USA (Tables 5–7). In Triana, Alabama, USA, levels were found to be equal to or to exceed those in many occupational exposures; and there were also high levels of exposure to polychlorinated biphenyls (Kreiss et al., 1981).

*Evidence of toxicity*

At present, there appears to be little controversy regarding the following points:
(1) DDT induces hepatomas in experimental animals at levels which may occur in humans.
(2) It is a well-recognized enzyme inducer, which is believed to cause tumours in animals in the same manner as phenobarbital. It has been shown to be an enzyme inducer in man (WHO, 1979).
(3) Experimental work supports the view that DDT acts as a promoter or enhancer in experimental models. In most of the large-scale experiments in which it has been tested, the mice or rats occasionally developed spontaneous liver tumours, so that the possibility of spontaneous initiation, with DDT acting as a promoter, cannot be excluded, as suggested by Pitot and Sirica (1980).

DDT levels such as those found in Triana are suspected of causing subtle changes in liver function, as suggested by the accelerated metabolism of some drugs in DDT workers as a result of enzyme induction. Associations with increased levels of triglycerides and serum cholesterol have also been reported. None of these effects has been associated with significant illness at Triana and elsewhere, but they suggest that the liver may be a primary target organ.

*Evidence for carcinogenicity in humans*

The epidemiological studies carried out in man have been summarized adequately elsewhere (Mrak, 1969; IARC, 1974; WHO, 1979; IARC, 1982), and these references

Table 4. Average DDT contents (ppm total DDT) in females in South Africa and Israel by age[a]

| Female subjects | Age group | | | | | | |
|---|---|---|---|---|---|---|---|
| | Months | Years | | | | | |
| | 0–11 | 1–4 | 5–14 | 15–24 | 25–44 | 45–69 | 70+ |
| South Africa | | | | | | | |
| White | – | 3.46 | 3.29 | 7.73 | 19.72 | 6.46 | – |
| Black | – | 5.85 | 2.76 | 4.73 | 8.22 | 7.87 | 11.37 |
| Israel | 5.34 | – | 18.68 | 14.33 | 15.93 | 19.39 | 20.92 |

[a] From IARC (1970)

Table 5. Temporal changes in geometric means of residues in adipose tissue of selected organochlorine pesticides in the US white population during the period 1970–1978[a]

| Year | No. of samples | Total DDT (ppm) | β-BHC[b] (ppm) | Dieldrin (ppm) |
|---|---|---|---|---|
| 1970 | 1 373 | 7.46 | 0.44 | 0.22 |
| 1971 | 1 545 | 7.04 | 0.37 | 0.24 |
| 1972 | 1 877 | 6.23 | 0.30 | 0.21 |
| 1973 | 1 085 | 5.70 | 0.32 | 0.21 |
| 1974 | 892 | 4.96 | 0.23 | 0.18 |
| 1975 | 771 | 4.67 | 0.24 | 0.16 |
| 1976 | 667 | 4.38 | 0.25 | 0.14 |
| 1977 | 775 | 3.35 | 0.20 | 0.12 |
| 1978 | 802 | 3.64 | 0.19 | 0.12 |

[a] F.W. Kutz, personal communication; by courtesy of J.E. Davies
[b] β-BHC, β-hexachlorocyclohexane

should be consulted for further details. Such studies have, in general, been directed to small groups of workers exposed to very high levels in occupational settings, to correlations of tissue levels with cancer at autopsy, or to correlations with cancer patterns in population groups exposed to a range of levels. The conclusions drawn by the IARC (1974) were: 'The cross-sectional epidemiological studies on workers exposed to DDT and the observational studies on volunteers were too limited and/or too short to allow any conclusions to be made regarding carcinogenesis.

'Although fat concentration of DDT residues were higher in terminal cancer patients than in control patients, this finding is inconclusive as to a causal relationship. A similar study with a different design did not show such a result.' Another working group (IARC, 1982) reported that although four studies had shown that 'tissue levels of DDT were ... higher in cancer patients than in subjects dying from other causes; no difference was found in two other studies.... Serum DDT appeared to be elevated in another study of nine cancer patients, but it is difficult to interpret. In two case-control studies of soft-tissue sarcoma and one of lymphoma, relative risks for the association of these diseases with exposure to DDT were, respectively, 1.2, 1.3

Table 6. Temporal changes in geometric means of residues in adipose tissue of selected organochlorine pesticides in the US black population during the period 1970–1978[a]

| Year | No. of samples | Total DDT (ppm) | β-BHC[b] (ppm) | Dieldrin (ppm) |
|---|---|---|---|---|
| 1970 | 1 373 | 12.97 | 0.69 | 0.35 |
| 1971 | 1 545 | 16.21 | 0.65 | 0.35 |
| 1972 | 1 877 | 12.16 | 0.47 | 0.27 |
| 1973 | 1 085 | 9.66 | 0.37 | 0.27 |
| 1974 | 892 | 8.44 | 0.31 | 0.19 |
| 1975 | 771 | 7.04 | 0.27 | 0.16 |
| 1976 | 667 | 6.70 | 0.30 | 0.16 |
| 1977 | 775 | 4.47 | 0.19 | 0.11 |
| 1978 | 802 | 6.07 | 0.23 | 0.14 |

[a] F.W. Kutz, personal communication; by courtesy of J.E. Davies
[b] β-BHC, β-hexachlorocyclohexane

Table 7. Geometric means and ranges of persistent pesticides (total DDT) in fat of populations in Caribbean countries and in the USA[a]

| Country | Population | Years | No. of Samples | Positive (%) | Geometric mean (ppm) | Range (ppm) |
|---|---|---|---|---|---|---|
| The Biminis (Bahamas) | General | 1971 | 98 | 100 | 58.5 | 33.6–381.1 |
| Haiti | Migrants | 1981–1982 | 54 | 100 | 106.2 | 24.7–655.9 |
| Jamaica | General | 1981 | 84 | 98 | 10.9 | 2.1– 98.2 |
| Trinidad and Tobago | Occupationally exposed | 1981 | 108 | 100 | 38.2 | 16.0– 72.0 |
| Trinidad and Tobago | General | 1981–1982 | 51 | 82 | 23.4 | 5.4–236.1 |
| USA | General | 1975–1980 | 308 | 98.7 | 10.4 | 0 – 80.9 |

[a] J.E. Davies, personal communication

and 1.6. Some of the men had also been exposed to phenoxyacetic acids and chlorophenols, which gave higher relative risks. A case-control study of colon cancer showed no increased relative risk for exposure to DDT. A small excess of deaths from cancer (3 observed, 1.0 expected) was found in forestry foremen exposed to DDT; 2,4-D; and 2,4,5-T.' In summarizing the data, the group concluded that DDT should be placed in Group 2B, i.e., that it is probably carcinogenic to humans.

In the absence of any definitive positive epidemiological study and in view of the impossibility of proving a negative, any conclusion concerning this type of data must rely on subjective evaluation and on experience and reasoned judgement. Although the conclusions of the two IARC working groups are essentially similar, they may lead to different perceptions by those members of the public and the scientific community who are unfamiliar with the data. The *IARC Monographs* have tended historically to use a conservative and prudent approach to evaluation. In attempting to evaluate the strength of negative evidence, some monographs, however, do not

Table 8. Concentration of DDT (DDT and DDE)[a] in fat and cancer rates per 100 000 per annum[b]

| Country | Years | Total DDT (ppm) | Cancer rates | | | |
|---|---|---|---|---|---|---|
| | | | Liver | Female breast | Female leukaemia[c] | Female lymphoma |
| UK | 1965–1967 | 3.0 | 1.0 | 53.0 | 4.3 | 3.6 |
| Denmark[d] | 1965 | 3.3 | 4.7 | 49.1 | 3.9 | 4.0 |
| Germany, Federal Republic of | 1958–1959 | 2.3 | 3.6 | 48.4 | 3.7 | 3.9 |
| German Democratic Republic | 1966–1967 | 13.1 | 4.5 | 33.4 | 4.5 | 3.9 |
| Poland | 1965 | 13.4 | 8.5 (urban) 4.2 (rural) | 31.5 | 5.0 | 1.7 |
| Hungary | 1960 | 12.4 | 0.9 | 19.8 | 2.3 | 1.3 |
| Spain | 1966 | 15.7 | 7.8 | 30.6 | 3.3 | 2.4 |
| Israel | 1965–1966 | 18.1 | 2.5 | 55.5 | 6.1 | 9.7 |
| India | 1964 | 26.0 | 1.4 | 20.1 | 2.3 | 1.6 |
| New Zealand | 1965–1969 | 14.6 | 1.9 | 52.5 | 5.1 | 3.6 |

[a] From Davies and Edmundson (1972)
[b] From Waterhouse et al. (1982)
[c] ICD 200 & 201
[d] ICD 7

appear to me to have given adequate weight to available descriptive epidemiological studies, nor have they discussed the biological consistency between suspected mechanisms and potential target organs in humans. Accordingly, it is worthwhile to give some consideration here to the information or inferences that can be drawn from correlations. It is widely recognized that epidemiological correlations have relatively limited value in demonstrating causation, although they may suggest hypotheses; however, in certain situations, they may provide circumstantial evidence to support the view that if no effect can be demonstrated, the impact in humans may be insignificant or at least small, especially when exposures in total populations have been considerable and prolonged (see also Doll, p. 6).

In addition to causing hepatoma and hepatocellular carcinoma in animals, DDT is reported to lead to increased incidences of leukaemias and lymphomas. By extrapolating to humans and assuming that the mechanisms in animals are not dissimilar qualitatively from those in humans, it could be anticipated that DDT might be a promoting agent for those human tumours in which promoting activity is believed to be important, such as those of the liver, breast, prostate and endometrium.

Utilizing the data in *Cancer Incidence in Five Continents Vol. IV* (Waterhouse et al., 1982) and information on existing DDT levels (Table 8), the incidence rates for cancer in a number of countries are presented, with data on total DDT levels in fat tissues. It is obvious that there is no correlation between DDT levels and cancer incidence at the sites selected. In Singapore and Israel, various ethnic subgroups show very different cancer patterns, although population exposures to DDT are probably similar. Further, where differences in incidence are observed, other, more plausible explanations are available. It should be recognized, of course, that such general correlations can exclude only relatively strong effects; however, these inferences are consistent and are supported by other circumstantial evidence, as well as by the limited case-control studies.

If DDT operates in humans in a way similar to that in animals, the type of tumour that should be observed would be a hepatoma arising as a nodular hyperplasia similar to that reported with steroid hormones. There are no data in the literature to show any general increase in this tumour that could be related to widespread DDT exposure, in spite of the fact that such tumours have been studied rather extensively due to their relationship with the use of steroid hormones (IARC, 1979). If DDT is related to tumours of the liver arising in cirrhosis, there is again no evidence that any change in liver cancer incidence can be correlated with DDT usage. If DDT had had a significant impact relative to other cirrhogenic factors, I think it should have shown up, since cirrhosis is becoming an increasingly recognized disease in many countries, although the rates of liver carcinoma in many areas, such as North America, have been falling slightly, partly due to diagnostic reasons.

Similar arguments can be made regarding data from Africa, the Middle East and Asia, where the use of DDT has been very extensive, e.g., in Israel and Japan. There again, there has been no evidence of an increase in liver cancer. In Africa and southeast Asia, much better arguments can be made that the changes in incidence are related to exposures to aflatoxin and/or hepatitis B virus. It is argued, however, that the exposures have been too low to lead to nodular or diffuse liver hyperplasia, even though, in many cases, exposures were very high and similar to those in animal studies.

In examining the possible role of DDT in the causation of other tumours, I am unable to detect any widespread pattern in changes in incidence that would in any way parallel increasing exposures to a pesticide, possibly acting as a promoter, between 1950 and 1970, followed by a fall in exposure. This applies equally to cancers of the breast, endometrium, prostate and other cancers in which enzyme induction might be important. Unfortunately, the available case-history studies are inadequate for studying soft-tissue tumours, since these tumours are rare and show minimal geographical and time-trend variations.

In certain countries, notably the USA, there is evidence that there is some difference in metabolism between blacks and whites and among social classes. There is no evidence that cancer distribution patterns with respect to social class in the USA are consistent with the use of DDT. If the levels of cancer in the north and south of the USA are examined (Table 9), they do not suggest any correlation with DDT levels.

The points made above are consistent with the report of the WHO (1979), which also emphasized the failure to show any evidence of a relationship between liver or other cancers and DDT. It also pointed out the lower frequency of liver cancer in rural areas than in metropolitan areas in general, which is in contrast to the distribution of DDT in tissues. The WHO report pointed out that DDT levels are very high in many different organs and that it would be difficult to explain cancer in any of these organs on the basis of such deposits alone.

Table 9. Concentrations of DDT in fat[a] in the US population and cancer rates per 100 000 per annum

| Area[b] | Race | Total DDT (ppm) | Cancer rates | | | |
|---|---|---|---|---|---|---|
| | | | Liver | Female breast | Female leukaemia | Female lymphoma |
| North | White | 4.84 | 3.0 | 29.5 | 4.9 | 5.9 |
| | Black | 7.86 | 5.1 | 27.7 | 4.1 | 2.9 |
| South | White | 9.5 | 4.6 | 28.7 | 5.2 | 5.5 |
| | Black | 14.33 | 4.1 | 30.3 | 7.0 | 3.2 |

[a] From Davies and Edmundson (1972)
[b] Cancer rates for the North are from Detroit; those for the South are from New Orleans

## Other considerations

It has been reported that there may be more chromatid breaks in people using DDT, but such workers have been exposed to many other insecticides (WHO, 1979). The limitations of such cytological studies are well-recognized, especially when they are based on cultured lymphocytes.

With regard to the role of enzyme induction, none of the chemicals in Group 1 in the *IARC Monographs Supplement 4* (IARC, 1982), apart from steroids, can be considered as non-genotoxic, although many carcinogens are noted inducers. Obviously, much more research is needed on the dose-responses of DDT and other enzyme inducers to define the potential role of these and similar compounds in human carcinogenicity, if any. Pitot and Sirica (1980) have suggested that a number of agents, such as phenobarbital, act by promoting background initiated foci and are not in themselves true initiators. As of yet, no tumour of the liver in man has been ascribed to the use of phenobarbital (see p. 153), which is believed to operate in a way similar to DDT, although in some experiments DDT is a weaker promoter than phenobarbital. In view of the observed incidence of liver tumours in certain mice administered DDT, it is obvious that there must be significant differences between mice and humans in their reactions to similar doses of DDT, suggesting that the mouse is a much more susceptible species.

## CONCLUSIONS

The *IARC Monographs Supplement 4* (IARC, 1982) concludes that the human data are inadequate to demonstrate that DDT is a carcinogen for humans. Unfortunately, this statement may imply for some people that there is a real but so far undetected effect. The preamble to the *Monographs* does not appear to address adequately the question of whether the weight of the epidemiological evidence suggests that an effect may be so small that, for all practical purposes, it cannot be detected nor is likely ever to be detected, in humans. Thus, the evaluation of negative evidence must remain a matter of judgement based on all data.

All the available evidence in humans indicates that adenomas of the liver, or benign hepatomas, are exceedingly rare. Reported cases are usually related to the use of oral contraceptives. The demonstration of a small number of such tumours in women was sufficient to identify the problem and to indicate the causal agent; there has subsequently been increased interest in these tumours in women. If DDT, at the levels used, had had an effect of such magnitude on the liver, the increase would almost certainly have been identified. So it is probable that DDT has had much less impact than oestrogens in relation to the female liver. This is of particular interest in view of the fact that in some countries with well-developed pathology laboratories, which can identify and study rare tumours, study population subgroups that have had considerable variations in exposure have been available for investigation.

Similar arguments can be used for tumours at other sites; but, again, the impossibility of proving a negative never permits the definitive conclusion that a substance is without effect. However, if DDT has an effect on human cancer, it must be exceedingly small. It is unlikely that further meaningful information will be garnered, although defined high-exposure groups should be followed even if they are exposed to multiple pesticides. A negative result would lend further support to these conclusions and would suggest that the various pesticides do not potentiate each other under the usual circumstances of occupational exposure.

In conclusion, therefore, in evaluating DDT, the fact that it is probably a late-stage carcinogenic agent in animals and, thus, its effects can be seen relatively soon should be taken into consideration, if we are to give any credence to our knowledge of animal models in terms of mechanisms and assume that they may be of value in demonstrating potential risk in humans. Secondly, the available case-control studies in high-risk groups, although limited, have shown nothing. If descriptive epidemiological data in humans are of value, the available evidence regarding cancer patterns and DDT distribution suggests that if DDT is an 'operational' carcinogen in humans, it is a very weak one. Thus, I conclude that the available data provide no evidence that DDT has been demonstrated to be a significant risk to humans at the levels to which they have been exposed in the past.

## REFERENCES

Davies, J.E. & Edmundson, W.F., eds (1972) *Epidemiology of DDT,* Mount Kisco, NY, Futura Publishing Co.

Davies, J.E., Edmundson, W.F., Raffonelli, A., Cassady, J.C. & Morgade, C. (1972) The role of social class in human pesticide pollution. *Am. J. Epidemiol., 96,* 334–341

IARC (1970) *Annual Report, 1970,* Lyon

IARC (1974) *IARC Monographs on the Evaluation of Carcinogenic Risk of Chemicals to Man,* Vol. 5, *Some Organochlorine Pesticides,* Lyon, pp. 83–124

IARC (1979) *IARC Monographs on the Evaluation of the Carcinogenic Risk of Chemicals to Humans,* Vol. 21, *Sex Hormones (II),* Lyon

IARC (1982) *IARC Monographs on the Evaluation of the Carcinogenic Risk of Chemicals to Humans,* Suppl 4, *Chemicals, Industrial Processes and Industries Associated with Cancer in Humans (IARC Monographs, Volumes 1 to 29),* Lyon

Kreiss, K., Zack, M.M., Kimbrough, R.D., Needham, L.L., Smrek, A.L. & Jones, B.T. (1981) Cross-sectional study of a community with exceptional exposure to DDT. *J. Am. med. Assoc., 245,* 1926–1930

Mrak, E.M. (1969) *Report of the Secretary's Commission on Pesticides and Their Relationship to Environmental Health,* Parts I and II, Washington DC, US Government Printing Office

Pitot, H.C. & Sirica, A.E. (1980) The stages of initiation and promotion in hepatocarcinogenesis. *Biochim. biophys. Acta, 605,* 191–215

Waterhouse, J., Shanmugaratnam, K., Muir, C., Powell, J., Peacham, D. & Whelan, S., eds (1982) *Cancer Incidence in Five Continents, Vol. IV (IARC Scientific Publications No. 42),* Lyon

WHO (1979) *Environmental Health Criteria 9: DDT and Its Derivatives,* Geneva

# DDT:

# CONCLUSION

### Rapporteur: T.W. ANDERSON

The classification of DDT as '2B' by the IARC implies that the 'chemical (concerned), group of chemicals, industrial process or occupational exposure is probably carcinogenic to humans'. The data from studies on experimental animals require the Agency to place DDT in group 2, but the weakness of the human evidence, which was considered as inadequate, required its separation from the small group of chemicals for which there was 'at least limited evidence of carcinogenity to humans', which are classified as '1A'.

The experimental data suggest that the liver would be the organ most likely to be affected by DDT. Data on liver cancer in humans over the age of 60 is unreliable, since the majority of cases are likely to be secondary rather than primary. It is therefore advisable when comparing liver cancer mortality in different countries to restrict comparison to people of age 55 or less. In a current study of oral contraceptives, the liver cancer rates in the UK, the USA, Australia and the Federal Republic of Germany are very similar in people below age 55 and seem to be unchanged since the early 1950s. There is thus no evidence from these data that the widespread use of DDT since the 1940s has caused any measurable increase in liver cancer in humans.

The relevance of animal experiments indicating an increase in liver cancer was questioned. Evidence is accumulating that DDT is similar to a large number of other compounds that increase the frequency of liver tumours in certain strains of mice; and it is probably unjustified to assume that they all cause cancer in man. In the case of DDT, this over-interpretation of toxicological data may have resulted in a substantial number of human deaths from insect-borne diseases that would otherwise have been prevented. Thus, the concern over DDT has probably been responsible for an increase rather than a decrease in human mortality.

Useful studies of the incidence of liver cancer in humans in western countries are almost impossible to carry out because of the overwhelming effect of alcohol. The incidence of liver cancer has been shown to be increased in women taking oral contraceptives, but, since males and females have probably been exposed to DDT to a similar extent (if anything, males even more than females), the absence of any increase in males in recent years argues strongly in favour of DDT being non-carcinogenic in humans in the concentrations normally encountered.

Case-control studies of the level of DDT in the body fat of cancer patients are confounded by the effect of cachexia, which is likely to result in an increased tissue concentration of DDT; it is therefore probably an artefact resulting from the cancer, rather than reflecting a high exposure to DDT in earlier life. Also, some of the case-control studies that appear to implicate DDT as a carcinogen must be interpreted cautiously because of the possible effect of other pesticides to which the patients were exposed.

It was suggested that this problem could be avoided if DDT levels were measured in cancer patients when the cancer was first diagnosed, before any cachexia had occurred. However there are large variations in DDT levels in different body tissues and in different individuals. One would therefore need a very large series for a study to be meaningful. Furthermore, the cost of obtaining tissue samples for analysis of DDT is very high. Thus, in recent work in the USA, the cost of preparing a sample was in the region of $250, not including the cost of the analysis. It was also pointed out that any large-scale, expensive study that involves a follow-up of 5, 10 or even 20 years is extremely difficult to carry out because of the need to have renewed funding at frequent intervals.

It was felt that, in view of the fact that large amounts of DDT were manufactured in several countries over a period of 10 to 15 years, it should be possible to identify persons involved in its manufacture and to follow them up in terms of cancer mortality. Unfortunately, those companies manufacturing DDT were also involved in manufacturing other pesticides, so that it was almost impossible to find a pure DDT exposure among production workers. However, it was pointed out that this would be a problem only if there were indeed an increased incidence of cancer in these workers. On the one hand, a negative result would exonerate DDT, whether or not other pesticides were involved. On the other hand, a positive result would of course be difficult to interpret because of the possible effect of other pesticides.

Because of its low cost, it appears that DDT will continue to be manufactured and used in the Third World. It would therefore be possible to establish information on workers producing DDT in those countries and to follow them over a period of years. Unfortunately, there would be a substantial delay before any conclusion could be drawn. It therefore seemed more sensible to take a hard look at existing data on persons who were exposed in the manufacturing process in the 1940s and 1950s.

Measurement of DDT in human organs is complicated by the fact that organs store DDT at different levels. Thus, body fat, which is commonly used as the reference tissue for human studies, tends to contain levels of DDT that are some ten times those typically found in the liver.

At first sight, pesticide applicators would appear to be an obvious group for follow-up study, but, unfortunately, it appears in most situations that sprayers are a relatively transient population, and poor records have usually been kept of their exposure and subsequent health experience. A recent follow-up study on some 16 000 pesticide applicators in the USA has given essentially negative results. (There had been one death from liver cancer and about one was expected.) The relevance of this study of DDT is uncertain, since the sprayers were identified back to 1967, and the number that were employed before that time was very small.

It was pointed out that even if DDT could be shown to be perfectly safe for humans, it was unlikely that it would be permitted again in the USA or other western countries because of environmental concerns. Furthermore, the body burdens of DDT in some human groups is extremely high and its persistence in human tissue makes it a substance that most regulators would treat with caution.

There is a paucity of sound epidemiological evidence concerning DDT. Its use has been severely restricted in many parts of the world on the basis of evidence of carcinogenesis in animals (the significance of which may be questioned) and of environmental effects on sensitive species. Its persistence in human tissues following exposure adds to the concern of possible toxicity. In comparison with other pesticides, it is relatively cheap to manufacture and therefore will probably continue to be used in some parts of the world. In some situations, even if there is a slight risk of long-term harm to humans from the use of DDT, it is outweighed by the short-term benefits of control of insect vectors of disease.

As with most carcinogenic agents, the long latent period of induction of cancer makes new prospective studies rather unattractive. It was felt generally that a vigorous effort was therefore justified in attempting to identify groups of persons exposed in the past to high levels of DDT. These include in particular those individuals involved in its manufacture, since applicators of pesticides tend to be exposed only intermittently and to be less easy to identify and follow up.

There may also be some groups of individuals who were exposed on a non-occupational basis and who are worth following closely. Such a group exists at Triana, Alabama, USA where, because of poorly controlled disposal of waste materials, the burden of DDT in the tissues of local inhabitants is extremely high (higher in many cases than those typically found in persons involved in its manufacture).

Even though DDT use has been discontinued in many western countries it is important to pursue the question of its long-term toxicity to humans, since, because of its low cost, it is potentially of great value in the Third World, and it is likely that the net benefit to humanity would be positive, even if it did indeed have a weak carcinogenic effect.

The evidence that DDT is carcinogenic to man is inadequate to permit a firm conclusion, but suggests that it is unlikely to have produced a quantitatively large increase in risk under the conditions of exposure that have operated in the past.

# SACCHARIN/CYCLAMATES

# SACCHARIN/CYCLAMATES;

# LABORATORY EVIDENCE

### P. SHUBIK
*Green College, Oxford, UK*

*Saccharin*

Sodium saccharin presents a unique problem to workers in experimental carcinogenesis. High doses can give rise to bladder cancers in male rats when exposure is begun either *in utero* or at a very early age. It does not appear to have a similar effect under other conditions in rats, or in hamsters or mice. It is difficult to compare some of these negative and positive studies because of the many differences in doses and other experimental parameters; however, when saccharin is tested by the standard techniques on which regulation of food additives is based no adverse effect is found. In three studies in which saccharin was fed to two generations of Sprague-Dawley rats (Tisdel *et al.*, 1974; Arnold *et al.*, 1977, 1980; WHO, 1984), up to 30% of the male $F_1$ rats developed bladder tumours, many of which were transitional-cell carcinomas.

In a more recent large-scale study, efforts were made to obtain a dose-response for this effect: 2500 male $F_1$ Sprague-Dawley rats fed doses of 1, 3, 5, 6.25 or 7.5% sodium saccharin for two generations were investigated for bladder tumour induction. The group sizes were designed to be fitted to certain mathematical models that have been used in attempting carcinogenic risk assessments and were 350 untreated controls, 700 fed 1%, 500 at 3%, 200 at 4% and 125 each at 5%, 6.25% and 7.5%. Bladder tumours were seen in animals at the 3% level, and the incidence steadily increased in groups up to the 7.5% levels (Table 1). Of particular interest is a group of 125 rats fed sodium saccharin starting at birth (initially through the mother's milk) and then fed 5% in the diet; this group developed almost as many tumours as those 125 rats the parents of which were fed sodium saccharin prior to mating and throughout gestation. In a prior study from the Health Protection Branch in Ottawa, it had been suggested that male rats fed sodium saccharin starting at 37 days of age were susceptible to bladder carcinogenesis – although far fewer tumours occurred than in those animals fed throughout gestation. Clearly the precise timing of this occurrence needs to be established.

Following the original studies of Hicks *et al.* (1975), a variety of investigations have been made of the effects of saccharin on rats initially treated with a bladder

Table 1. Percentage incidence (and number) of rats with primary bladder neoplasms

| Dietary concentration (%) | No. of bladders examined | Transitional-cell papilloma, polyp | Transitional-cell carcinoma | Total |
|---|---|---|---|---|
| *Sodium saccharin* | | | | |
| 0.00 | 324 | 0.0 (0) | 0.0 (0) | 0.0 (0) |
| 1.00 | 658 | 0.6 (4) | 0.2 (1) | 0.8 (5) |
| 3.00 | 472 | 0.8 (4) | 0.8 (4) | 1.7 (8) |
| 4.00 | 189 | 2.1 (4) | 4.2 (8) | 6.3 (12) |
| 5.00 | 120 | 3.3 (4) | 9.2 (11) | 12.5 (15) |
| 6.25 | 120 | 10.0 (12) | 6.7 (8) | 16.7 (20) |
| 7.50 | 118 | 15.3 (18) | 16.1 (19) | 31.4 (37) |
| 5.00 (throughout gestation) | 122 | 0.0 (0) | 0.0 (0) | 0.0 (0) |
| 5.00 (following gestation) | 120 | 3.3 (4) | 6.7 (8) | 10.0 (12) |
| *Sodium hippurate* | | | | |
| 3.00 | 118 | 0.0 (0) | 0.0 (0) | 0.0 (0) |
| *Historical controls* | | | | |
| (10 studies) | 863 | 0.7 (6) | 0.1 (1) | 0.8 (7) |

carcinogen. Chowaniec and Hicks (1979) fed rats with sodium saccharin following intravesicular instillation of $N$-nitroso-$N$-methylurea; the studies of Ito (Nakanishi *et al.*, 1980) involved the carcinogen $N$-butyl-$N$-(4-hydroxybutyl)nitrosamine and those of Cohen *et al.* (1979), $N$-[4-(5-nitro-2-furyl)-2-thiazolyl]formamide (FANFT). In all of these studies, a secondary effect, called 'promotion', has been said to have been demonstrated. However, the individual studies differ widely and must be considered in great detail before a general conclusion is arrived at.

Sodium saccharin is non-mutagenic in submammalian assays; however, it has been found to be capable of damaging chromosomes in other assays (IARC, 1980). It does not appear to be metabolized, and extensive studies have failed to shake this conclusion. The most recent biochemical studies demonstrate that saccharin can modify the metabolism of tryptophan, resulting in the presence of increased quantities of indoxyl in the urine.

Sodium saccharin can, therefore, induce bladder cancer in male Sprague-Dawley rats under specific conditions of administration. Effectively large doses—3% and upwards—are required to produce a significant result. In the most recent large-scale study (WHO, 1984), the dose-response observed ruled out the 'one-hit' model but was consistent with Weibull and multihit models. In this large-scale study, although a significantly increased incidence of tumours occurred in animals fed 1%, attention should be drawn to the fact that five tumours occurred among the 700 rats in this group whereas none was seen among the 350 untreated controls. The meaning of this occurrence is clearly debatable.

*Cyclamates*

Sodium cyclamate was originally tested (Price *et al.*, 1970) chronically in rats in combination with sodium saccharin as a 10:1 mixture, since this was the form in which it was used in humans. Later in this study (Oser et al., 1975), the protocol was amended to add cyclohexylamine to the diet. Three dose levels (500, 1120 or 2500 mg/kg) were used, and 9/25 male and 3/25 female rats developed transitional-cell tumours of the bladder.

The only outstandingly different result was obtained by Hicks et al. (1975) who fed cyclamate at either 1.0 or 2.0 g/kg body weight to rats following a single initial intravesicular administration of the powerful, locally acting carcinogen $N$-nitroso-$N$-methylurea. Over 50% of the rats developed bladder tumours quickly. When $N$-nitroso-$N$-methylurea was given alone no tumour was seen, and with cyclamate alone a 1% incidence of bladder tumours occurred. Repetition of this study by Mohr et al. (1978) did not confirm these findings.

There are now few scientists who believe that cyclamate is a bladder carcinogen. Clearly, the first study is impossible to interpret, since, in addition to the fact that three compounds were given, the rats bore parasites.

A major remaining problem does not concern carcinogenesis but rather the testicular atrophy caused by the metabolite of cyclamate – cyclohexylamine. This amine is produced through the action of gut flora in variable amounts at various times, since both rats and humans initially have a certain number of 'convertors', and these may increase in number as bacteria mutate. Recent unpublished work by Renwick has shown that cyclohexylamine is metabolized differently in rats and humans, and the significance of extrapolation is therefore questionable.

Like saccharin, cyclamate does not appear to cause mutations, but cyclohexylamine has an effect in some tests (IARC, 1980).

## REFERENCES

Arnold, D.L., Moodie, C.A., Stavrić, B., Stoltz, D.R., Grice, H.C. & Munro, I.C. (1977) Canadian saccharin study. *Science, 197,* 320

Arnold, D.L., Moodie, C.A., Grice, H.C., Charbonneau, S.M., Stavrić, B., Collins, B.T., McGuire, P.F. & Munro, I.C. (1980) Long term toxicity of *ortho*toluenesulfonamide and saccharin in the rat. *Toxicol. appl. Pharmacol., 52,* 113–152

Chowaniec, J. & Hicks, R.M. (1979) Response of the rat to saccharin with particular reference to the urinary bladder. *Br. J. Cancer, 39,* 355–375

Cohen, S.M., Arai, M., Jacobs, J.B. & Friedell, G.H. (1979) Promoting effect of saccharin and DL-tryptophan in urinary bladder carcinogenesis. *Cancer Res., 39,* 1207–1217

Hicks, R.M., Wakefield, J.St.J. & Chowaniec, J. (1975) Evaluation of a new model to detect bladder carcinogens or co-carcinogens; results obtained with saccharin, cyclamate and cyclophosphamide. *Chem.-biol. Interactions, 11,* 225–233

IARC (1980) *IARC Monographs on the Evaluation of the Carcinogenic Risk of Chemicals to Humans,* Vol. 22, *Some Non-nutritive Sweetening Agents,* Lyon

Mohr, U., Green, U., Althoff, J. & Schneider, P. (1978) *Syncarcinogenic action of saccharin and sodium-cyclamate in the induction of bladder tumours in MNU-*

*pretreated rats.* In: Guggenheim, B., ed., *Health and Sugar Substitutes,* Basel, Karger, pp. 64–69

Nakanishi, K., Hirose, M., Ogiso, T., Hasegawa, R., Arai, M. & Ito, N. (1980) Effects of sodium saccharin and caffeine in the urinary bladder of rats treated with $N$-butyl-$N$-(4-hydroxybutyl)nitrosamine. *Gann, 71,* 490–500

Oser, B.L., Carson, S., Cox, G.E., Vogin, E.E. & Sternberg, S.S. (1975) Chronic toxicity study of cyclamate:saccharin (10:1) in rats. *Toxicology, 4,* 315–330

Price, J.M., Biava, C.G., Oser, B.L., Vogin, E.E., Steinfeld, J. & Ley, H.L. (1970) Bladder tumors in rats fed cyclohexylamine or high doses of a mixture of cyclamate and sodium. *Science, 167,* 1131–1132

Tisdel, M.O., Nees, P.O., Harris, D.L. & Derse, P.H. (1974) *Long-term feeding of saccharin in rats.* In: Inglett, G.E., ed., *Symposium: Sweeteners,* Westport, CN, Avi Publishing Co., pp. 145–148

WHO (1984) Evaluation of certain food additives and contaminants. 28th Report of the Joint FAO/WHO Expert Committee on Food Additives. *Tech. Rep. Ser. 710,* Geneva

# SACCHARIN/CYCLAMATES:

# EPIDEMIOLOGICAL EVIDENCE

### B.K. ARMSTRONG

*NH & MRC Research Unit in Epidemiology and Preventive Medicine,*
*University of Western Australia, Nedlands, Western Australia*

## SUMMARY

Adequate data on the carcinogenicity of saccharin and cyclamate to humans are available only for the urinary bladder. In the studies available, exposure to saccharin and to cyclamate cannot be distinguished readily.

Descriptive studies have shown no evidence of time trends in bladder cancer that can be related to use of saccharin or cyclamate. Likewise, studies of diabetics, who have used more saccharin and cyclamate than other people, have shown no evidence of an increased risk of bladder cancer. This association, however, is probably confounded negatively by cigarette smoking.

Thirteen case-control studies have addressed the relationship of saccharin and cyclamate intake to bladder cancer in individuals. While statistically significant positive associations have been observed, a similar number of significant negative associations has also been observed. Studies of the dose-response relationships have also shown no consistent pattern. Studies of saccharin and cyclamate use with smoking habits have shown no consistent interaction with heavy smoking, as might be expected from a promotional effect. In some studies, however, an increased risk with saccharin and cyclamate use has been observed in female non-smokers—a group otherwise at low risk for bladder cancer.

## INTRODUCTION

Saccharin was first synthesised in 1879 and entered commercial use in a limited way in the USA and Europe around the turn of the century. Its use increased with restrictions on availability of sugar during the First World War and again during the Second. Its increase in use has, however, been largely a post-Second World War phenomenon.

Cyclamate is of more recent origin. It was discovered in 1944 and introduced commercially as an artificial sweetener in 1952. The main increase in its use occurred between 1960 and 1969, when it was banned in the USA and in some European countries; however, it continues to be used in many other countries (e.g., Brazil, Federal Republic of Germany, Finland, Pakistan, South Africa, Switzerland) (IARC, 1980).

Of the saccharin used in the USA, about 45% is taken in diet soft drinks, 18% in table-top sweeteners, 23% in other food preparations and the rest in miscellaneous cosmetics, pharmaceuticals and tobacco preparations. Before it was banned in the USA, cyclamate had a similar distribution of uses (IARC, 1980).

In 1972 in the UK, estimated intake of saccharin was 8 g per person per year (Armstrong & Doll, 1975). In 1969, before the ban on cyclamate in the USA, estimated consumption of cyclamate was 44 g per person per year.

Consideration of saccharin and cyclamates as specific causes of human cancer is made difficult by the fact that, for a decade at least (from 1960 to 1969), in most countries, saccharin and cyclamates were commonly used as a 1:10 mixture (1 saccharin:10 cyclamate). While attempts have been made in some epidemiological studies to separate saccharin and cyclamate exposure, it is doubtful whether this is possible. In most studies, therefore, particularly those involving recalled individual exposure, exposure must be considered as mixed. The term 'artificial sweetener' has been used in this review to refer to saccharin, cyclamate or their mixture, except when saccharin or cyclamate was specifically identified. Saccharin has, of course, been used in substantial quantities since the Second World War and has continued to be used following bans on the use of cyclamate, in a number of countries. Cyclamate was in common use for only a decade, at least in the USA, Canada, the UK and a number of other European countries. Most of the results may be taken, therefore, as referring mainly to saccharin, although measures of 'dose' of saccharin may be inaccurate for the period during which cyclamate was in common use.

The principal hypothesis generated by experimental studies is that artificial sweeteners (particularly saccharin) cause bladder cancer. Epidemiological studies of artificial sweeteners have, therefore, concentrated on bladder cancer. Some studies have, in addition, examined the relationship between artificial sweetener use and other cancers (Armstrong et al., 1976; Morrison, 1979). These studies have not suggested any association between artificial sweetener use and cancer in general or any specific cancer. In the presence of other evidence that artificial sweeteners cause a particular cancer other than bladder cancer, however, it is doubtful that those studies could be considered as sufficient to establish a negative. The rest of this review is confined to bladder cancer.

## DESCRIPTIVE STUDIES

For the purpose of this review, a descriptive study is defined as a study in which incidence or mortality of bladder cancer is related to the aggregate artificial sweetener consumption of a whole population. There are six such studies—three relating trends in artificial sweetener use over time to trends in bladder cancer and three comparing cancer rate in a population with high average artificial sweetener use (diabetics) with that in the general population.

With respect to time trends, Burbank and Fraumeni (1970) related bladder cancer mortality in the USA, and bladder cancer incidence in Connecticut between 1950 and 1967, to cyclamate use between 1950 and 1969. (Cyclamate use increased 10-fold between 1960 and 1969.) There was no apparent association between the two. Armstrong and Doll (1974) related bladder cancer mortality in England and Wales from 1911 to 1970 to saccharin use in the period 1939 to 1972; again, there was no apparent association. More recently, Jensen and Kamby (1982) compared bladder cancer incidence from 1961 to 1976 in Danes born in 1941–1945, a period in which saccharin use in Denmark reached very high levels, with that in Danes of similar ages born between 1931 and 1940 (before the wartime rise in saccharin use). There was no evidence that those born in 1941–1945, presumed to have had high in-utero exposure to saccharin, had an increased subsequent risk of bladder cancer.

The studies of diabetics (Kessler, 1970; Armstrong & Doll, 1975; Armstrong *et al.*, 1976) have likewise given negative results. Diabetics have, if anything, a reduced rather than an increased risk of bladder cancer, which is accompanied by reduced risks of other smoking-related cancers.

While reassuring at the time and suggestive that artificial sweeteners had not increased risk of bladder cancer to a major degree, these studies are limited in a number of respects. No association would have appeared in the studies of time trends, even if artificial sweeteners did cause bladder cancer, if the minimum latent period from first exposure to clinical cancer was greater than about seven years (in the study of cyclamate) or 30 years (in the study of saccharin); if the effects were comparatively small; or if other trends were present (as may indeed have been the case for cigarette-induced bladder cancer in the UK) that obscured any effect of artificial sweeteners. Similar considerations apply to the Danish study of in-utero exposure; in addition, no effect would have been expected in this study if long-continued post-natal exposure were required before cancers would appear.

The studies of diabetics are also limited. No effect may have been evident if it were confined to a comparatively small number of heavy users of artificial sweeteners. In addition, the possible confounding effects of other risk factors, particularly cigarette smoking, could not be controlled.

These limitations on interpretation of the descriptive studies mean that, in the presence of other evidence suggestive of weak carcinogenic effects of artificial sweeteners in humans, these studies cannot be taken as providing evidence to the contrary. The final conclusion depends, therefore, on the results of the analytical epidemiological studies.

## ANALYTICAL STUDIES

The essential difference between analytical and descriptive epidemiological studies is that analytical studies document the exposure and disease status of individual people.

Thirteen case-control studies have related individual use of artificial sweeteners to risk of bladder cancer. Details of these studies, together with results of dichotomous exposure analyses (usually 'ever use' or 'never use' of artificial sweeteners, whether 'table-top' use, in diet drinks, or other use) are given in Table 1. Their quality varies; in only one, however, does it seem likely that bias has seriously affected the results. In that study (Morgan & Jain, 1974), the female control subjects were all patients with stress incontinence. It would be reasonable to postulate that these patients would have been more likely than most to use artificial sweeteners – thus explaining the significantly low relative risk for bladder cancer (0.35). None of the studies offered any validation of their measurement of artificial sweetener use.

In two studies, risk of bladder cancer was increased significantly in relation to some measure of artificial sweetener use. Wynder and Goldsmith (1977) (see footnote 14 in Wynder and Stellman, 1980) observed a relative risk of 1.43 (95% confidence interval, 1.10–1.86) with regular use of artificial sweeteners in 402 male cases and 7200 male controls adjusted for age, hospital, hospital status and year of interview. Howe et al. (1977) found a relative risk of 1.6 (95% confidence interval, 1.1–2.3) for 'ever use' of artificial sweeteners in 480 male case-control pairs matched for age and neighbourhood. Two significantly negative relationships were found, however (in addition to that of Morgan and Jain, 1974, referred to above) between 'table-top' artificial sweetener use and bladder cancer, in 223 male and 66 female cases from Nagoya, Japan (Morrison et al., 1982). There is no obvious methodological explanation for these low relative risks, although it is noted that most cases in this study were interviewed in hospital (in-patient or out-patient), whereas most controls were interviewed at home. Age and smoking history were controlled in the analysis. In summary, there were four statistically significant results (not obviously due to bias), two positive and two negative, in a total of 47 separate estimates of relative risk of bladder cancer in association with some two-category measure of exposure to artificial sweeteners.

Seven of the 13 studies referred to above described trends in relative risk of bladder cancer in relation to increasing amount, frequency or duration of use of 'table-top' artificial sweeteners or diet drinks. The details are shown in Table 2. Statistically significant trends (two-tailed test) were observed in three studies. Relative risk increased significantly with increasing average frequency and duration of use of 'table-top' artificial sweeteners in males in the study of Howe et al. (1977) and with increasing duration of use and in females in the study of Morrison and Buring (1980). Relative risk of bladder cancer *decreased* significantly with increasing duration of use of 'table-top' artificial sweeteners in males in Nagoya, in the study of Morrison et al. (1982). These three significant trends arose in 40 separate trend analyses *reported* from the seven studies [for this count, the sexes, exposure categories ('table-top', diet drinks, other) and amount/frequency and duration were considered as separate].

Although there was no apparent trend in relative risks with increasing use in separate analyses, the joint distribution of frequency of daily use of 'table-top' artificial sweeteners and diet drinks in the study of Hoover and Strasser (1980) showed the highest relative risk in those who used 'table-top' sweeteners more than six times daily and diet drinks twice or more daily (relative risk, 1.45 with 95% confidence

Table 1. Case-control studies relating bladder cancer risk to use of artificial sweeteners

| Reference | Period | Cases | | | Controls | | | Matched | Variables controlled | Data collection | Definition of exposure | Relative risks | | | | | |
|---|---|---|---|---|---|---|---|---|---|---|---|---|---|---|---|---|---|
| | | | | | | | | | | | | Males | | | Females | | |
| | | Source | No. | % responding[a] | Source | No. | % responding[a] | | | | | Sugar substitute | Diet drink | Other | Sugar substitute | Diet drink | Other |
| Morgan & Jain (1974) | NS[b] | NS | 232 | 64 | NS | 232 | 49 | Yes | Age | Mail questionnaire | Prolonged regular use | – | – | 1.0[c] (0.6–1.8) | – | – | 0.4[c] (0.2–0.8) |
| Simon et al. (1975) | 1965–1971 | 10 hospitals in Massachusetts and Rhode Island; lower urinary-tract cancer; 95% bladder | 134 | <62 | 10 hospitals, excl. patients with urinary problems | 382 | 59 | Yes | Age, residence, year of discharge | Mail questionnaire | Current use of saccharin in tea or coffee | – | – | – | 1.0 (0.5–1.7) | – | – |
| Wynder & Goldsmith (1977); Wynder & Stellman (1980) | 1973–1977 | 17 hospitals in 6 US cities | 524 | >96[d] | 17 hospitals, excl. patients with tobacco-related disease | 10 977 | >96 | No / Yes | Age, hospital, hospital status, year Above, plus education | Interview | Regular use | 1.4 (1.1–1.9) / 1.1 (0.6–2.1) | – | – | 0.9 (0.5–1.6) / 0.8 (0.2–3.0) | – | – |
| Howe et al. (1977); Miller & Howe (1977); Howe et al. (1980) | 1974–1976 | All new cases in British Columbia, Nova Scotia and Newfoundland | 632 | 77 | Neighbourhood and electoral roll | 632 | 72 | Yes | Age, neighbourhood | Interview | Ever use | 1.6 (1.1–2.3) | 0.8 (0.2–3.3) | 1.2 (0.6–2.3) | 0.6 (0.3–1.1) | 0.9 (0.2–3.0) | 0.5 (0.3–1.0) |
| Kessler & Clark (1978) | 1972–1975 | 19 hospitals in Baltimore area | 519 | 45 | 19 hospitals; excl. patients with cancer and bladder conditions | 519 | ≃68 | Yes | Age, race, marital status, time of admission | Interview | More than occasionally | 0.9 (0.6–1.2) | 1.0 (0.6–1.4) | 1.2 (0.8–1.9) | 0.9 (0.6–1.5) | 1.0 (0.6–1.6) | 0.7 (0.4–1.3) |
| Miller (1978) | NS | Urology out-patient clinic | 265 | NS | Other urology and other patients | 530 | NS | Yes | Age | Self admin. quest. | Used regularly | 1.1[e] | – | – | 0.9[e] | – | – |
| Connolly et al. (1978) | NS | NS | 348 | NS | NS | 696 | NS | Yes | Age, area of residence | NS | Ever use | 0.9[f] (0.6–1.4) | – | – | 0.7[f] (0.4–1.3) | – | – |
| Wynder & Stellmann (1980) | 1977–1979 | US hospitals; new cases | 397 | NS | US hospitals; other conditions | 397 | NS | Yes | Age, hospital, hospital room status | Interview | Ever used | 0.9[g] (0.7–1.3) | 0.8 (0.6–1.2) | – | 0.6[g] (0.3–1.4) | 0.6 (0.3–1.3) | – |

Table 1 continued.

| Reference | Period | Cases | | | Controls | | | Matched | Variables controlled | Data collection | Definition of exposure | Relative risks | | | | | | |
|---|---|---|---|---|---|---|---|---|---|---|---|---|---|---|---|---|---|---|
| | | | | | | | | | | | | Males | | | | Females | | |
| | | Source | No. | % responding[a] | Source | No. | % responding[a] | | | | | Sugar substitute | Diet drink | Other | | Sugar substitute | Diet drink | Other |
| Hoover & Strasser (1980) | 1978 | All new cases in 10 defined areas of US | 3 479 | 86 | Random sample of total population of same areas | 5 784 | ≃73 | No | Age, race, cigarettes, coffee, occupation, diabetes, area, education | Interview | Ever used | 1.0 (0.9–1.2) | 1.0 (0.8–1.1) | 1.0 (0.8–1.2) | | 1.0 (0.8–1.3) | 1.0 (0.8–1.2) | 1.1 (0.9–1.5) |
| Morrison & Buring (1980) | 1976–1977 | All Boston area hospitals | 592 | 80 | Random sample of Boston residents | 536 | 79 | No | Age | Interview | Ever used | 0.8 (0.5–1.1) | 0.8 (0.6–1.1) | – | | 1.5 (0.9–2.6) | 1.6 (0.9–2.7) | – |
| Cartwright et al. (1981) | NS | Prevalent and incident cases, Yorkshire | 841 | NS | Other hospital patients without cancers | 1 060 | NS | Partially (analysis not matched) | Age, type of case | Interview | Saccharin taker | 1.2[h] (0.9–1.7) | – | – | | 1.3[h] (0.8–2.3) | – | – |
| Morrison et al. (1982) | 1976–1978 | All hospitals in Manchester UK and Nagoya, Japan; new cases | 839 | 91 | Electoral registers | 1 552 | 84 | No | Age, smoking history | Interview | Ever used | M[f] 0.9 (0.7–1.2)<br>N[f] 0.7 (0.5–0.9) | 0.9 (0.5–1.6) | 1.0 (0.4–2.1) | | 0.9 (0.6–1.4)<br>0.5 (0.3–0.8) | 0.9 (0.4–1.8) | 1.3 (0.6–2.8) |
| Najem et al. (1982) | 1978 | Four clinics and 2 hospitals, New Jersey | 150 | NS | Other patients of same clinics and hospitals excl. cancer or tobacco-related diseases | 76 | NS | Yes | Age, birthplace, race, source of case, residence | Interview | Regularly used | | | | | 1.3[i] (0.6–2.8) | 1.2[i] (0.6–2.1) | – |

[a] Wherever possible, % analysed of all those *ascertained* is quoted.
[b] NS, not stated
[c] All artificial sweeteners
[d] 'Refusal rate' stated to be <4%
[e] 95% confidence interval not stated and insufficient data given for its calculation
[f] Based on distributions of cases and controls, not discordant triplets
[g] Includes diet foods
[h] Combines results shown separately for smokers and non-smokers in table IV of Cartwright et al. (1981). Thus, relative risks are crude.
[f] M, Manchester; N, Nagoya
[i] Both sexes combined

Table 2. Results of trend analyses of risk by amount/frequency and duration of use of artificial sweeteners in case-control studies providing relevant data

| Reference | Variables controlled | Definition of exposure | Exposure level | Males RR | 95% CI | $\chi_1^2$ (trend) | Females RR | 95% CI | $\chi_1^2$ (trend) |
|---|---|---|---|---|---|---|---|---|---|
| Wynder & Goldsmith (1977) | Age, hospital, hospital status, year matched Matching broken | Duration of regular use of artificial sweeteners | Non-use<br><5 years<br>5-14 years<br>≥15 years | 1.0[a]<br>0.8<br>0.6<br>0.9 | (0.3–2.2)<br>(0.2–2.2)<br>(0.1–14.8) | 0.6 | 1.0[a]<br>0.6<br>0.9<br>0.9 | (0.1–3.9)<br>(0.1–15.4)<br>(0.1–15.4) | 0.0 |
| Howe at el. (1977); Miller & Howe, (1977); Howe et al. (1980) | Age, race, marital status, time of admission matched Matching broken | Average frequency of use of artificial sweeteners | Never used<br><2 500 tabs/yr<br>≥2 500 tabs/yr | 1.0<br>1.5<br>2.1 | (1.0–2.5)<br>(0.9–5.1) | 5.6 | | | |
| | | Duration of use | <3 years<br>≥3 years | 1.4<br>2.0 | (0.9–2.5)<br>(1.2–4.0) | 4.7 | | | |
| | | Frequency x duration | <3 y; <2 500<br>≥3 y; <2 500<br><3 y; ≥2 500<br>≥3 y; ≥2 500 | 1.6<br>1.6<br>1.3<br>5.3 | (0.8–2.9)<br>(0.8–3.3)<br>(0.4–4.2)<br>(1.4–20.8) | NS[b] | | | |
| | Smoking, occupation, water supply, bladder infection, diabetes, education, aspirin use. Logistic model | Average frequency per day of artificial sweetener use | 1-4/day<br>5-6/day<br>7-8/day<br>9+/day | 0.9<br>1.6<br>1.1<br>2.8 | (0.4–2.1)<br>(0.6–4.3)<br>(0.3–4.0)<br>(0.9–8.9) | NS | 0.3<br>0.5<br>1.2 | (0.1–1.1)<br>(0.1–2.9)<br>(0.2–5.3) | NS |
| Kessler & Clark (1978) | Smoking, occupation, age, race, diabetes, marital status education, obesity, dieting, memory | Amount and frequency of use of any artificial sweetener | Non-user<br>Low<br>Medium<br>High | 1.0<br>1.1<br>1.4<br>1.0 | (0.7–1.8)<br>(0.8–2.3)<br>(0.6–1.7) | NS | 1.0<br>1.2<br>0.7<br>0.7 | (0.6–2.4)<br>(0.3–1.8)<br>(0.3–1.5) | NS |

[a] Based on the 132 male cases, 124 male controls, 31 females cases and 29 female controls reported in 1977
[b] NS, not stated

Table 2 continued.

| Reference | Variables controlled | Definition of exposure | Exposure level | Males | | | Females | | |
|---|---|---|---|---|---|---|---|---|---|
| | | | | RR | 95% CI | $\chi_1^2$ (trend) | RR | 95% CI | $\chi_1^2$ (trend) |
| Wynder & Stellman (1980) | Age, hospital, hospital status, year, education matched. Matching broken | Units of artificial sweetener ($\simeq$ 20–40 mg saccharin) | None<br>$\leq$1/day<br>2-3/day<br>4+/day | 1.0<br>1.2<br>1.1<br>0.6 | (0.7–2.2)<br>(0.6–1.9)<br>(0.4–1.2) | 0.7 | 1.0<br>0.4<br>0.9<br>1.1 | (0.1–1.2)<br>(0.2–3.8)<br>(0.3–4.5) | 0.1 |
| | | Units of diet drink (equal to 12-oz can) | $\leq$3/week<br>4–14/week<br>15+/week | 1.2<br>0.6<br>1.2 | (0.6–2.3)<br>(0.3–1.1)<br>(0.4–3.9) | 0.9 | 0.6<br>0.5 | (0.2–1.8)<br>(0.1–1.8) | 1.7 |
| | | Duration of artificial sweetener use | 0–4 years<br>5–10 years<br>11+ years | 1.0<br>0.9<br>1.1 | (0.6–1.7)<br>(0.5–1.5)<br>(0.5–2.4) | 0.0 | 0.7<br>1.1<br>0.3 | (0.2–2.1)<br>(0.3–3.8)<br>(0.1–1.4) | 1.4 |
| | | Duration of diet drink use | 0–4 years<br>5–10 years<br>11+ years | 0.8<br>0.8<br>0.9 | (0.4–1.6)<br>(0.4–1.6)<br>(0.3–2.3) | 0.5 | 0.4<br>0.5<br>1.8 | (0.1–1.4)<br>(0.1–2.3)<br>(0.2–19.9) | 0.6 |
| Hoover & Strasser (1980)[c] | Age, race, smoking | Average daily uses of 'table-top' artificial sweeteners | Never used<br><1<br>1–1.9<br>2–3.9<br>4–5.9<br>$\geq$6 | 1.0<br>1.1<br>0.9<br>1.1<br>1.0<br>1.0 | NS<br>NS<br>NS<br>NS<br>NS | 0.2 | 1.0<br>0.7<br>1.3<br>1.4<br>1.0<br>1.4 | NS<br>NS<br>NS<br>NS<br>NS | 1.9 |
| | | Average daily use of diet drinks | <1<br>1–1.9<br>2–2.9<br>$\geq$3 | 0.9<br>0.9<br>1.4<br>1.0 | NS<br>NS<br>NS<br>NS | 0.4 | 1.0<br>0.8<br>1.7<br>1.4 | NS<br>NS<br>NS<br>NS | 0.9 |
| | | Years of use of 'table-top' artificial sweeteners[d] | <5<br>5–9<br>$\geq$10 | 1.0<br>1.0<br>1.0 | NS<br>NS<br>NS | 0.1 | 1.0<br>1.4<br>1.0 | NS<br>NS<br>NS | 0.7 |
| | | Years of use of diet drinks[d] | <5<br>5–9<br>10–14<br>$\geq$15 | 1.0<br>1.0<br>1.0<br>0.8 | NS<br>NS<br>NS<br>NS | 1.4 | 1.1<br>1.1<br>1.1<br>0.8 | NS<br>NS<br>NS<br>NS | 0.2 |

[c] Joint distribution of number of uses of 'table-top' sweeteners by number of uses of diet drinks in males and females together showed highest risk in the three heaviest use categories (RRs, 1.53, 1.56, 1.64)

[d] From Hoover et al unpublished data

Table 2 continued.

| Reference | Variables controlled | Definition of exposure | Exposure level | Males RR | 95% CI | $\chi^2_1$ (trend) | Females RR | 95% CI | $\chi^2_1$ (trend) |
|---|---|---|---|---|---|---|---|---|---|
| Morrison & Buring (1980) | Age, smoking | Artificial sweetener powder/ packets per day | No use<br><3<br>3+ | 1.0<br>0.5<br>1.0 | <br>NS<br>NS | 1.3[e] | 1.0<br>1.2<br>1.3 | <br>NS<br>NS | 0.8[e] |
| | | Tablets per day | <5<br>5+ | 1.5<br>1.3 | NS<br>NS | 0.0[e] | NS<br>NS | | |
| | | Diet drinks/day | <1<br>1<br>2+ | 0.7<br>0.9<br>1.9 | NS<br>NS<br>NS | 0.0[e] | 2.5<br>1.6<br>0.5 | NS<br>NS<br>NS | 0.3[e] |
| | | Diet foods served/week | <1<br>1-2<br>3+ | 0.8<br>0.6<br>0.9 | NS<br>NS<br>NS | 1.4[e] | 0.5<br><br>1.4 | NS<br><br>NS | 0.2[e] |
| | | Duration of use of artificial sweeteners | <5 yr<br>5-9 yr<br>10+ yr | 0.7<br>0.8<br>1.0 | NS<br>NS<br>NS | 0.7 | 1.0<br><br>3.7 | NS<br><br>NS | 6.5[e] |
| | | Duration of use of diet drinks | <5 yr<br>5-9 yr<br>10+ yr | 1.1<br>0.6<br>0.7 | NS<br>NS<br>NS | 2.5[e] | 1.3<br><br>1.3 | NS<br><br>NS | 1.5[e] |
| Morrison et al. (1982) | Age, smoking | Frequency of use of artificial sweetener tablets (Manchester) | No use<br><5<br>5-9 yr<br>10+ yr | 1.0<br>0.8<br>0.8<br>0.6 | <br>NS<br>NS<br>NS | 0.8[e] | 1.0<br>0.7<br>0.6<br>2.3 | <br>NS<br>NS<br>NS | 0.0[e] |
| | | Duration of use of artificial sweeteners (Manchester) | <3 yr<br>3-5 yr<br>6-8 yr<br>9-14 yr<br>15+ yr | 0.7<br>0.9<br>1.6<br>0.3<br>0.9 | NS<br>NS<br>NS<br>NS<br>NS | 0.0[e] | 1.3<br>0.5<br>1.2<br>0.4<br>0.9 | NS<br>NS<br>NS<br>NS<br>NS | 1.1[e] |
| | | (Nagoya) | <3 yr<br>3-5 yr<br>6-8 yr<br>9+ yr | 0.6<br>0.8<br>0.7<br>0.5 | NS<br>NS<br>NS<br>NS | 4.0[e] | 0.4<br>0.5<br><br>0.6 | NS<br>NS<br><br>NS | 2.8[e] |

[e] Values of chi-squared (trend) were not given for adjusted relative risks; these values of $\chi^2_1$ are not adjusted for confounders.

interval, 1.00 to 2.10, adjusted for sex, age, race, smoking, occupation, region and education).

Where duration of use has been studied, some 2-4% of subjects (cases and controls) had taken saccharin for more than about 15 years.

Hoover and Strasser (1980) decided *a priori* to examine relative risk in a low-risk group—female non-smokers without employment in a hazardous occupation—and in a high-risk group—male smokers of > 40 cigarettes per day; the former to display any absolute effect of artificial sweeteners more easily (because it would be seen against a lower background risk) and the latter to display any co-carcinogenic, presumably promotional, effect. In the event, both groups showed trends toward increasing risk of bladder cancer with increasing frequency of use of artificial sweeteners, although neither trend was statistically significant at conventional levels with a two-tailed test.

Table 3 summarizes all available data relevant to effect modification by prior risk of bladder cancer. In most studies, cigarette smoking was the relevant 'risk' variable. With respect to female non-smokers, data are available from three studies in addition to that of Hoover and Strasser (1980). Morrison and Buring (1980) found the highest risk in female non-smokers for both 'ever use' of artificial sweeteners (relative risk, 2.1) and 'ever use' of diet drinks (relative risk, 2.6); they did not provide confidence intervals for these age-adjusted relative risks. Similarly, Cartwright *et al.* (1981) found the highest risk in female non-smokers (relative risk, 1.6 with 95% confidence interval, 0.8 to 3.2) and male non-smokers (relative risk, 2.2 with 95% confidence interval, 1.3 to 3.8). Morrison *et al.* (1982) found the highest relative risks in male and female non-smokers in Manchester (1.6 in males and 1.2 in females) but not in Nagoya. The relative risk in female non-smokers was probably less than unity in the study of Kessler and Clark (1978; see footnote b to table 3).

With respect to male smokers, there is no support for the hypothesis of Hoover and Strasser, except in the study of Howe *et al.* (1977). In this study, relative risk of bladder cancer with 'ever use' of artificial sweeteners was 1.7 (95% confidence interval, 0.8 to 3.4) in male smokers of ≥ 15 cigarettes per day but also 2.1 (with 95% confidence interval, 1.1 to 3.9) in male ex-smokers.

In examining this effect modification by prior risk of bladder cancer, Walker *et al.* (1982) reanalysed the data of Hoover and Strasser (1980) using a risk score which included the contributions to risk from education, history of bladder infection, occupational exposure to carcinogens, coffee consumption and cigarette smoking. There was essentially no evidence of effect modification. Hoover and Hartge (1982), however, have pointed out that inclusion of both males and females in the single analysis significantly dilutes the degree of risk (or lack of it) at the extremes of the distribution. Thus, this approach was inefficient in testing the hypothesis.

The results of these various analyses are further summarized in Table 4. The counts given ignore questions of study size, study quality and statistical significance. For analyses using a dichotomous definition of exposure, studies that might be interpreted as showing a *protective* effect outnumber those that might be interpreted as showing a *causal* effect. For analyses of amount/frequency of use or duration of use at more than two exposure levels, the numbers suggesting causal and protective trends are very nearly equal. With respect to effect modification by smoking, studies showing

Table 3. Summary of risk of bladder cancer with artificial sweetener use in 'low-risk' and 'high-risk' groups

| Reference | Variables controlled | Definition of exposure | Risk strata | Males RR | Males 95% CI | Females RR | Females 95% CI |
|---|---|---|---|---|---|---|---|
| Howe et al. (1977); Miller & Howe (1977); Howe et al. (1980) | None | 'Ever use' of artificial sweeteners | Never smoked | 0.7 | (0.2–1.3) | NS[a] | — |
| | | | Ex-smoker | 2.1 | (1.1–3.9) | NS | — |
| | | | <15 cigs/day | 1.0 | (0.3–3.3) | NS | — |
| | | | ≥15 cigs/day | 1.7 | (0.8–3.4) | NS | — |
| Kessler & Clark (1978) | None | Use of 'table-top' artificial sweeteners | Non smokers | | | 1.4[b] | — |
| | | | Smokers | | | 0.8 | — |
| Wynder & Stellman (1980) | Some; not stated which | Ever use' of artificial sweeteners | Current long-term (≥10 yr) cigarette smokers | 0.6 | (0.3–1.1) | 1.0 | (0.2–5.1) |
| Hoover & Strasser (1980) | Age | 'Ever use' of 'table-top' artificial sweeteners | Very low risk[c] | NS | | 3.0 | (0.9–9.2)[d] |
| | | | Low risk[c] | NS | | 1.2 | (0.9–1.7)[d] |
| | | 'Ever use' of diet drinks | Very low risk[c] | NS | | 1.1 | (0.4–4.8)[d] |
| | | | Low risk[c] | NS | | 1.1 | (0.7–1.4)[d] |
| | | 'Ever use' of 'table-top' artificial sweeteners | Never smoked | 0.9[e] | (0.7–1.2) | 1.2[e] | (0.9–1.7) |
| | | | <20 cigs/day | 1.1 | (0.9–1.3) | 0.9 | (0.6–1.3) |
| | | | 21–40 cigs/day | 0.8 | (0.6–1.1) | } 1.3 | (0.7–2.4) |
| | | | 40+ cigs/day | 1.7 | (1.1–2.6) | | |
| | | 'Ever use' of diet drinks | Never smoked | 0.8[e] | (0.6–1.0) | 0.9[e] | (0.7–1.2) |
| | | | <20 cigs/day | 1.2 | (0.9–1.4) | 0.9 | (0.7–1.3) |
| | | | 21–40 cigs/day | 0.8 | (0.6–1.0) | } 1.7 | (1.0–3.0) |
| | | | 40+ cigs/day | 1.4 | (0.9–2.0) | | |

[a] NS, not stated
[b] Unadjusted RR in male non-smokers was 1.69; adjusted for 'possible confounding factors' was 2.61 with 95% CI 1.20–5.67; RR among female non-smokers presumably less than unity.
[c] Very low, never smoked cigarettes, never drank coffee, no hazardous occupation; low, same but coffee drinkers not excluded. Data on 'very low risk women' from Howe et al., unpublished progress report.
[d] Confidence interval calculated from raw data, unadjusted for confounders; little difference between adjusted and unadjusted point estimates of RR
[e] RRs not adjusted for age as recalculated from raw data; age not strongly confounding. Some evidence of trend to increasing RRs with increasing frequency of use of artificial sweeteners or diet drinks in males who smoked 40+ cigarettes per day and females who smoked 20+ cigarettes per day

Table 3. Continued.

| Reference | Variables controlled | Definition of exposure | Risk strata | Males RR | 95% CI | Females RR | 95% CI |
|---|---|---|---|---|---|---|---|
| Walker et al. (1982) (independent analysis of data of Hoover & Strasser, 1980) | Age, sex, race, region | 75+ servings week-years of artificial sweetener use[f] | Risk score[a] 1 | | | 1.0 | (0.7–1.4) |
| | | | 2 | | | 0.9 | (0.6–1.4) |
| | | | 3 | | | 1.3 | (0.9–2.0) |
| | | | 4 | | | 1.2 | (0.8–1.9) |
| | | | 5 | | | 0.8 | (0.5–1.2) |
| Morrison & Buring (1980) | Age | 'Ever used' artificial sweeteners | Non-smoker | 1.1 | NS | 2.1 | NS |
| | | | Ex-smoker | 1.0 | NS | 1.7 | NS |
| | | | Current smoker | 0.6 | NS | 1.2 | NS |
| | | 'Ever used' diet drinks | Non-smoker | 0.9 | NS | 2.6 | NS |
| | | | Ex-smoker | 1.1 | NS | 1.9 | NS |
| | | | Current smoker | 0.7 | NS | 1.5 | NS |
| Cartwright et al. (1981) | Age, whether prevalent or incident case | 'Ever used' saccharin as an artificial sweetener | Non-smoker | 2.2 | (1.3–3.8) | 1.6 | (0.8–3.2) |
| | | | Smoker | 0.9 | (0.6–1.3) | 1.2 | (0.5–2.6) |
| Morrison et al. (1982) | Age | 'Ever used' artificial sweeteners | (Manchester) Non-smoker | 1.6 | NS | 1.2 | NS |
| | | | Ex-smoker | 0.9 | NS | 1.0 | NS |
| | | | Current smoker | 0.9 | NS | 0.9 | NS |
| | | | (Nagoya) Non-smoker | 0.5 | NS | 0.4 | NS |
| | | | Ex-smoker | 0.6 | NS | – | |
| | | | Current smoker | 0.7 | NS | 0.8 | NS |

[f] Highest exposure level defined; results similar at all other exposure levels
[a] Score derived from subjects' education, occupational exposure to bladder carcinogens, coffee consumption and cigarette smoking

Table 4. Summary of pertinent results from 13 case-control studies relating artificial sweetener use to bladder cancer

| Study result | Number of relevant studies | | Number of studies with this result | | | |
|---|---|---|---|---|---|---|
| | | | Males | | Females | |
| | 'Table top' | Diet drinks | 'Table top' | Diet drinks | 'Table top' | Diet drinks |
| *Two exposure levels ('ever used'/'never used')* | 11 | 6 | | | | |
| Risk increased by 25 % + | | | 2 | 0 | 2 | 1 |
| Risk decreased by 20% + | | | 2 | 3 | 4 | 1 |
| *More than two exposure levels (trend analyses)* | | | | | | |
| Amount and frequency of use | 6 | 3 | | | | |
|   Highest risk in highest exposure category | | | 1 | 1 | 4 | 0 |
|   Lowest risk in highest exposure category | | | 2 | 0 | 1 | 2 |
| Duration of use | 6 | 3 | | | | |
|   Highest risk in longest use category | | | 1 | 0 | 1 | 2 |
|   Lowest risk in longest use category | | | 1 | 1 | 1 | 1 |
| *Effect modification by smoking* | | | | | | |
| Female non-smokers | 5 | 2 | | | | |
|   Risk increased by 25% + | | | – | – | 2 | 1 |
|   Risk decreased by 20% + | | | – | – | 1 | 0 |
| Male smokers | 7 | 2 | | | | |
|   Risk increased by 25% + | | | 2 | 1 | – | – |
|   Risk decreased by 20% + | | | 3 | 1 | – | – |

increased and decreased risks in male smokers are about evenly balanced, but there is a majority in favour of an increased risk in female non-smokers (three 'for' and one 'against').

## CONCLUSIONS

In the face of 13 analytical studies, at least some of which may be interpreted as positive, little weight can be given to the negative results of the descriptive studies. With respect to these 13 studies, the balance of evidence, based purely on numbers of studies, is against the hypothesis that use of artificial sweeteners is associated with an increased risk of bladder cancer in humans. No particular weight, however, has been given to study size and quality. In this regard, the study of Hoover and Strasser (1980) is both the largest and the best. Overall, it showed no evidence of increased or decreased risk of bladder cancer with use of any form of artificial sweetener in people of either sex. There was, however, evidence of an increased risk of marginal statistical significance in heavy users of both 'table-top' artificial sweeteners and diet drinks (relative risk, 1.45). Relative risk was also increased, but not significantly, in low-risk women and high-risk men. The effect in low-risk women gains some support from several other studies.

On the basis of the present evidence either of two conclusions is possible:
(1) Artificial sweeteners do not cause human bladder cancer, or
(2) Artificial sweeteners are a very weak cause of human bladder cancer, their effect being demonstrated best in those otherwise at low risk of the disease, i.e., the responsible agent appears to act as a cancer initiator.

The corollary to the second conclusion should be evaluated for plausibility against relevant experimental data.

## REFERENCES

Armstrong, B. & Doll, R. (1974) Bladder cancer mortality in England and Wales in relation to cigarette smoking and saccharin consumption. *Br. J. prev. soc. Med., 28*, 233–240

Armstrong B. & Doll, R. (1975) Bladder cancer mortality in diabetics in relation to saccharin consumption and smoking habits. *Br. J. prev. soc. Med., 29*, 73–81

Armstrong, B., Lea, A.J., Adelstein, A.M., Donovan, J.W., White, G.C. & Ruttle, S. (1975) Cancer mortality and saccharin consumption in diabetics. *Br. J. prev. soc. Med., 30*, 151–157

Burbank, F. & Fraumeni, J.R. (1970) Synthetic sweetener consumption and bladder cancer trends in the United States. *Nature, 227*, 296–297

Cartwright, R.A., Abid, R., Glashan, R. & Gray, B.K. (1981) The epidemiology of bladder cancer in West Yorkshire. A preliminary report on non-occupational aetiologies. *Carcinogenesis, 2*, 343–347

Connolly, J.G., Rider, W.D., Rosenbaum, L. & Chapman, J.A. (1978) Relation between the use of artificial sweeteners and bladder cancer. *Can. med. Assoc. J., 119*, 408

Hoover, R. & Hartge, P. (1982) Non-nutritive sweeteners and bladder cancer. *Am. J. publ. Health, 72*, 382–383

Hoover, R.N. & Strasser, P.H. (1980) Artificial sweeteners and human bladder cancer. *Lancet, i*, 837–840

Howe, G.R., Burch, J.P., Miller, A.B., Morrison, B., Gordon, P., Nelson, L., Chambers, L.W., Fodor, G. & Winsor, G.M. (1977) Artificial sweeteners and human bladder cancer. *Lancet, ii*, 578–581

Howe, G.R., Burch, J.D., Miller, A.B., Cook, G.M., Esteve, J., Morrison, B., Gordon, P., Chambers, L.W., Fodor, G. & Winsor, G.M. (1980) Tobacco use, occupation, coffee, various nutrients and bladder cancer. *J. natl Cancer Inst., 64*, 701–713

IARC (1980) *IARC Monographs on the Evaluation of the Carcinogenic Risk of Chemicals to Humans*, Vol. 22, *Some Non-nutritive Sweetening Agents*, Lyon, pp. 39–48, 63, 125

Jensen, O.M. & Kamby, C. (1982) Intra-uterine exposure to saccharin and risk of bladder cancer in man. *Int. J. Cancer, 29,* 507–509

Kessler, I.I. (1970) Cancer mortality among diabetics. *J. natl Cancer Inst., 44,* 673–686

Kessler, I.I. & Clark, J.P. (1978) Saccharin, cyclamate, and human bladder cancer. *J. Am. med. Assoc., 240,* 349–355

Miller, A.B. & Howe, G.R. (1977) Artificial sweeteners and bladder cancer. *Lancet, ii,* 1221–1222

Miller, C.T., Neutel, C.I., Nair, R.C., Marret, L.D., Last, J.M. & Collins, W.E. (1978) Relative importance of risk factors in bladder carcinogenesis. *J. chronic Dis., 31,* 51–56

Morgan, R.W. & Jain, M.G. (1974) Bladder cancer: smoking, beverages and artificial sweeteners. *Can. med. Assoc. J., 111,* 1067–1070

Morrison, A.S. (1979) Use of artificial sweeteners by cancer patients. *J. natl Cancer Inst., 62,* 1397–1399

Morrison, A.S. & Buring, J.E. (1980) Artificial sweeteners and cancer of the lower urinary tract. *New Engl. J. Med., 302,* 537–541

Morrison, A.S., Verhoek, W.G., Leck, I., Aoki, K., Ohno, Y. & Obala, K. (1982) Artificial sweeteners and bladder cancer in Manchester, UK, and Nagoya, Japan. *Br. J. Cancer, 45,* 332–336

Najem, G.R., Louria, D.B., Seebode, J.J., Thind, I.S., Prusakowski, J.M., Ambrose, R.B. & Fernicola, A.R. (1982) Life time occupation, smoking, caffeine, saccharine, hair dyes and bladder carcinogenesis. *Int. J. Epidemiol., 11,* 212–217

Simon, D., Yen, S. & Cole, P. (1975) Coffee drinking and cancer of the lower urinary tract. *J. natl Cancer Inst., 54,* 587–591

Walker, A.M., Dreyer, N.A., Friedlander, E., Loughlin, J., Rothman, K.J. & Kohn, H.I. (1982) An independent analysis of the National Cancer Institute study on non-nutritive sweeteners and bladder cancer. *Am. J. publ. Health, 72,* 376–381

Wynder, E.L. & Goldsmith, R. (1977) The epidemiology of bladder cancer. *Cancer, 40,* 1246–1268

Wynder, E.L. & Stellman, S.D. (1980) Artificial sweetener use and bladder cancer: A case control study. *Science, 207,* 1214–1216

SACCHARIN/CYCLAMATES:

# CONCLUSION

Rapporteur: D. KREWSKI

Of the methods currently available for assessing the carcinogenic potential of chemical substances, the most informative are epidemiological investigations of human populations and toxicological experiments using animal models for man (Office of Technology Assessment, 1981). While epidemiological studies are not subject to the uncertainties of trans-species extrapolation, other problems in interpretation can arise due to difficulties in accurately determining past levels of exposure and in adjusting for possible confounding factors. The protracted time scale of human carcinogenesis also presents serious problems in assessing the carcinogenic potential of substances to which humans have only recently been subjected at appreciable levels of exposure.

It must be emphasized that no study, toxicological or epidemiological, can provide absolute assurance of safety, regardless of the number of subjects involved (Krewski *et al.*, 1982; Day, 1985). While statistical sensitivity may be increased through the use of high dose levels in laboratory experiments, it is possible that this could invoke secondary mechanisms of carcinogenesis that may not be operative at lower levels of exposure (Clayson *et al.*, 1983).

Saccharin was first synthesized quite by accident over 100 years ago by Constantin Fahlberg while working in Ira Remsen's laboratory at Johns Hopkins' University in Baltimore (Remsen & Fahlberg, 1879–1880). Being many times sweeter than sugar, it came into widespread use as an inexpensive sugar substitute during shortages imposed by the First and Second World Wars. Cyclamate was discovered in 1944 and subsequently entered the marketplace as a second artificial sweetener. Mixtures of cyclamate:saccharin (10:1) were later introduced as a means of eliminating the unpleasant aftertaste of saccharin when used alone.

The controversy concerning the safety of saccharin is almost as old as saccharin itself (Bradshaw *et al.*, 1982). The most significant toxicological findings, however, are the bladder tumours induced in male Sprague-Dawley rats in two-generation feeding studies. Similar effects were noted by Arnold *et al.* (1980) following single-generation exposure. Recently, Fukushima *et al.* (1983) have reported positive results in ACI rats exposed for one year starting at six weeks of age, although the presence of the bladder thread worm *Trichosomoides crassicanda* in more than half of the animals complicates the interpretation of these findings.

Other studies using several initiating agents have suggested that saccharin may act as a promoter of bladder lesions in rats (see Arnold *et al.*, 1983, for detailed review). It produces inconsistent results in short-term tests for genotoxicity (Office of Technology Assessment, 1977; Arnold *et al.*, 1983), possibly due to impurities in the samples tested (Kramers, 1975). Saccharin is nucleophilic (Ashby *et al.*, 1978) and does not bind appreciably to DNA in rat liver or bladder (Lutz & Schlatter, 1977). Although apparently not metabolized itself, saccharin has been shown recently to modify tryptophan metabolism (Sims & Renwick, 1983), resulting in a greater formation of indole and excretion of urinary indican, two tryptophan metabolites known to have cocarcinogenic activity in the bladder.

Artificial sweeteners have been the subject of extensive epidemiological investigation, including temporal trend studies, studies of diabetics and case-control studies. While Howe *et al.* (1977) reported an overall increase in the risk of bladder cancer among males in their case-control study, any effect observed in other investigations has been limited to narrower subgroups of the study populations. Because of the difficulties in identifying small risks directly in exposed human populations, however, caution has been urged in the interpretation of these findings (Howe & Burch, 1981).

In a review of the data available at the time, the IARC (1982) considered the laboratory data on saccharin to provide limited evidence of carcinogenicity in animals, in that effects have been observed in only one sex of one species. This same review observed that the epidemiological studies showed no consistent evidence of an increased cancer risk among users of saccharin. However, cognizance was taken of an increasing trend in risk with both increasing consumption and duration of use observed in certain subroups of the Hoover and Strasser (1980) study. In the light of the animal evidence, these human data raised the possibility that saccharin may act as a weak carcinogen and/or promoter.

The case-control studies reported by Cartwright *et al.* (1981) and Morrison *et al.* (1982) since the time of the IARC review failed to produce any further evidence of carcinogenic effects in humans. A recently completed study in Copenhagen has also proved to be negative (Jensen *et al.*, 1983).

Taking this additional information into account, it was felt that the bulk of the epidemiological data is consistent with a lack of carcinogenic activity in humans. Because some small level of risk can never be ruled out on the basis of such data, however, it was allowed that saccharin may be a risk factor of unknown but probably minor consequence for human bladder cancer, but that any such effect was difficult to define accurately using current methods of epidemiological research. Although saccharin clearly can induce bladder cancer in rats, these effects are observed at relatively high doses administered continuously from birth throughout the majority of the animals' lifetime.

Additional toxicological experiments designed to elucidate the mechanism by which saccharin induces bladder lesions in Sprague-Dawley rats could be of value in assessing the relevance of the animal data to the human situation. One possible avenue of research would be an investigation of the effects of saccharin on the metabolism of amino acids other than tryptophan. Bioassay studies designed to clarify further the extent to which either perinatal or full lifetime exposure is required in the production of bladder tumours in rats could also be of value.

Further epidemiological investigations may have the greatest chance of success if focused on potential high-risk subgroups such as those suggested by Hoover and Strasser (1980), including non-smoking females with a low baseline risk. Knowledge of the mechanism by which saccharin induces rat bladder cancer may also facilitate subgroup selection. Any further investigation might also await the results of another completed but as yet unreported case-control study (Miller, personal communication).

Taken as a whole, the weight of the available human data was considered by the group to be consistent with the belief that saccharin, used either alone or in combination with cyclamates, is not a demonstrable cause of bladder cancer in the human population at large under past conditions of use. The effects observed in toxicological experiments occur under continuous exposure to relatively high dose levels from birth onwards. In the presence of laboratory evidence of limited generality the extensive epidemiological data base assembled to date was considered to weigh strongly against the possibility that saccharin is carcinogenic to man and, on balance, be of greater relevance to the human situation. Since the possibility of a weak effect cannot be completely excluded, however, more evidence on the risk of bladder cancer in non-smoking women in relation to the use of saccharin would be useful.

## REFERENCES

Arnold, D.L., Moodie, C.A., Grice, H.C., Charbonneau, S.M., Stavric, B., Collins, B.T., McGuire, P.F., Zawidzka, Z.Z. & Munro, I.C. (1980) Long-term toxicity of ortho-toluenesulfonamide and sodium saccharin in the rat. *Toxicol. appl. Pharmacol.* **52**, 113–152

Arnold, D.L., Krewski, D. & Munro, I.C. (1983) Saccharin: a toxicological and historical perspective. *Toxicology,* **27**, 179–256

Ashby, J., Styles, J.A., Anderson, D. & Paton, D. (1978) Saccharin: an epigenetic carcinogen/mutagen? *Food Cosmet. Toxicol.* **16**, 95–103

Bradshaw, L., Arnold, D.L. & Krewski, D. (1982) *A century of saccharin.* In: Burton, I., Fowle, C.D. & McCullough, R.S., eds, *Living With Risk: Environmental Risk Management in Canada (Environmental Monograph No. 3),* Toronto, Institute for Environmental Studies, University of Toronto, pp. 117–128

Cartwright, R.A., Adib, R., Glashan, R. & Gray, B.K. (1981) The epidemiology of bladder cancer in West Yorkshire. A preliminary report on non-occupational aetiologies. *Carcinogenesis,* **2**, 343–347

Clayson, D.B., Krewski, D. & Munro, I.C. (1983) The power and interpretation of the carcinogenicity bioassay. *Regul. Toxicol. Pharmacol.,* **3**, 329–348

Day, N.E. (1985) *Epidemiological methods for the assessment of human cancer risk.* In: Clayson, D., Krewski, D., & Munro, I. eds, *Toxicological Risk Assessment,* Boca Raton, FL, CRC Press (in press)

Fukushima, S., Ari, M., Nakanowatari, J., Hibino, T., Okuda, M. & Ito, N. (1983) Differences in susceptibility to sodium saccharin among various strains of rats and other animal species. *Gann, 74,* 8–20

Hoover, R.N. & Strasser, P.H. (1980) Artificial sweeteners and human bladder cancer. Preliminary results. *Lancet, i,* 837–840

Howe, G.R. & Burch, J.D. (1981) Artificial sweeteners in relation to the epidemiology of bladder cancer. *Nutr. Cancer, 2,* 213–216

Howe, G.R., Burch, J.D., Miller, A.B., Morrison, B., Gordon, P., Weldon, L., Chambers, L.W., Fodor, G. & Winsor, G.M. (1977) Artificial sweeteners and human bladder cancer. *Lancet, ii,* 578–581

IARC (1982) *IARC Monographs on the Evaluation of the Carcinogenic Risk of Chemicals to Humans,* Suppl. 4, *Chemicals, Indsutrial Processes and Industries Associated with Cancer in Humans (IARC Monographs, Volumes 1 to 29),* Lyon, pp. 224–226

Jensen, O.M., Knudson, J.B., Sorenson, B.L. & Clemmesen, J. (1983) Artificial sweeteners and absence of bladder cancer risk in Copenhagen. (Submitted for publication.)

Kramers, P.G.N. (1975) The mutagenicity of saccharin. *Mutat. Res., 32,* 81–92

Krewski, D., Clayson, D., Collins, B. & Munro, I.C. (1982) *Toxicological procedures for assessing the carcinogenic potential of agricultural chemicals.* In: Fleck, R.A. & Hollaender, A., eds, *Genetic Toxicology: An Agricultural Perspective,* New York, Plenum, pp. 461–497

Lutz, W.K. & Schlatter, Ch. (1977) Saccharin does not bind to DNA of liver or bladder in the rat. *Chem.-biol. Interact., 19,* 253–257

Morrison, A.S., Verhoek, W.G., Leck, I., Aoki, K., Ohno, Y. & Obata, K. (1982) Artificial sweeteners and bladder cancer in Manchester, UK and Nagoya, Japan. *Br. J. Cancer, 45,* 332–336

Office of Technology Assessment (1977) *Cancer Testing Technology and Saccharin,* Washington DC, US Government Printing Office

Office of Technology Assessment (1981) *Assessment of Technologies for Determining Cancer Risks from the Environment,* Washington DC, US Government Printing Office

Remsen, I. & Fahlberg, C. (1879–1880) On the oxidation of substitution products of aromatic hydrocarbons. IV. On the oxidation of orthotoluene sulphamide. *Am. chem. J., 1,* 426–438

Sims, J. & Renwick, A.G. (1983) The effects of saccharin on the metabolism of dietary tryptophan to indole, a known cocarcinogen for the urinary bladder of the rat. *Toxicol. appl. Pharmacol., 67,* 132–151

# PHENOBARBITAL

# PHENOBARBITAL:

# LABORATORY EVIDENCE

## J.R.P. CABRAL

*Unit of Mechanisms of Carcinogenicity, International Agency for Research on Cancer, Lyon, France*

Phenobarbital is a long-acting barbiturate, which has been used in medicine as a hypnotic, a sedative and in the treatment of epilepsy. The worldwide annual production of this compound is in the order of 100–1000 tonnes/year (IARC, 1977).

Short-term studies with phenobarbital were evaluated by IARC (1982) (Table 1) to be inadequate. Phenobarbital was not mutagenic in *Salmonella typhimurium*, and it did not elicit DNA repair in cultured rat liver cells. It was weakly mutagenic in *Drosophila melanogaster*. Phenobarbital did not induce cell transformation. No data on humans were available.

Phenobarbital has also been studied in long-term experiments in mice and rats; in all of these, the liver was the only target organ (Table 2). When given over the lifespan of rats (120 weeks), a concentration of 500 mg/l in drinking-water induced liver tumours in 32% of females and 59% of males, compared with a zero incidence in controls (Rossi *et al.*, 1977). No carcinogenicity study with phenobarbital in hamsters has been reported.

In initiation-promotion experiments, phenobarbital enhanced the incidence of liver tumours in mice previously treated with *N*-nitrosodimethylamine and in rats previously treated with *N*-2-acetylaminofluorene, *N*-nitrosodiethylamine, 4-dimethylaminobenzene or benzo[*a*]pyrene. However, when phenobarbital was given simultaneously with those compounds it reduced the incidence of liver tumours (IARC, 1977, 1982).

## REFERENCES

IARC (1977) *IARC Monographs on the Evaluation of Carcinogenic Risk of Chemicals to Man*, Vol. 13, *Some Miscellaneous Pharmaceutical Substances*, Lyon, pp. 157–181

Table 1. Phenobarbital: Evidence for activity in short-term tests[a]

|  | DNA damage | Mutation | Chromosomal anomalies | Other |
|---|---|---|---|---|
| Prokaryotes |  | − |  |  |
| Fungi/Green plants |  |  |  |  |
| Insects |  | ? |  |  |
| Mammalian cells (in vitro) | − |  |  | T(−)[b] |
| Mammals (in vivo) |  |  |  |  |
| Humans (in vivo) |  |  |  |  |

[a] From IARC (1982)
[b] T, cell transformation

Table 2. Experimental evidence of carcinogenicity of phenobarbital administered by the oral route[a]

| Species | No. of experiments | Dose (ppm) | Duration (weeks) | Results |
|---|---|---|---|---|
| Mouse | 3 | 500 | 52–120 | Liver tumours in both sexes |
| Rat | 2 | 500 | 64–120 | No excess of liver tumours in the 64-week experiment. Liver tumours in both sexes in the 120-week experiment |

[a] From IARC (1977)

IARC (1982) *IARC Monographs on the Evaluation of the Carcinogenic Risk of Chemicals to Humans,* Suppl. 4, *Chemicals, Industrial Processes and Industries Associated with Cancer in Humans (IARC Monographs, Volumes 1 to 29),* Lyon, pp. 208–211

Rossi, L., Ravera, M., Repetti, G. & Santi, L. (1977) Long-term administration of DDT or phenobarbital-Na in Wistar rats. *Int. J. Cancer,* **19,** 179–185

PHENOBARBITAL:

# EPIDEMIOLOGICAL EVIDENCE

B. MACMAHON

*Department of Epidemiology, Harvard School of Public Health, Boston, MA, USA*

Humans who receive the heaviest and most prolonged exposure to phenobarbital are probably patients with severe epilepsy, for whom it has been prescribed in daily doses of 100–300 mg—often over many decades. Two large groups of such patients have been studied.

In 1974, Clemmesen *et al.* reported a study of 9136 cases of epilepsy admitted to the Filadelfia treatment community in Denmark between 1933 and 1962. Cases of cancer occurring among them were identified by cross-matching against the nationwide cancer registry for the years 1943–1967. Apart from a striking excess of tumours of the central nervous system—ascribed by the authors to tumours for which epilepsy was the presenting symptom but which were latent at the time of diagnosis of epilepsy—there seemed, if anything, to be a deficit of tumours in this cohort. After excluding central nervous system tumours, 82 cancers were observed and 83.9 expected among epileptics treated for less than 10 years, but for those treated for more than 10 years there were 194.3 cancers expected and only 130 observed.

Clemmesen and Hjalgrim-Jensen (1977, 1978, 1980) later revised this series. They excluded 160 patients who were aliens, 131 who had died before follow-up began in 1943, 340 who were 'not traceable at all', 106 who had been previously registered twice because of data errors, and 321 in whom a diagnosis of epilepsy was not sustained. This left a cohort of 8078 individuals—4268 males and 3810 females. In addition to matching against the cancer registry, attempts were made to follow these subjects individually. For 98% of the cohort, follow-up could be done to 31 December 1972, or prior death. The follow-up of this series was subsequently extended through 1976 (with the loss of one additional male because of the discovery of double registration), and it is the report of this later follow-up that gives the most definitive picture of this series (Clemmesen & Hjalgrim-Jensen, 1981a). Four features are to be noted:

(1) The occurrence of central nervous system tumours declines sharply with number of years of observation (Table 1). The authors argue persuasively that this trend is the reverse of what would be expected if phenobarbital were causing brain tumours. Overall, I accept this argument, but I find it a little odd that nine cases of brain cancer,

Table 1. Observed and expected numbers of brain tumours by interval since admission for treatment. Sexes combined[a]

| No. of years since admission | No. of brain tumours | |
|---|---|---|
| | Observed | Expected |
| 10 | 45 | 3.9 |
| 10–14 | 13 | 2.4 |
| 15–19 | 4 | 2.3 |
| 20–24 | 5 | 1.9 |
| 25–29 | 2 | 1.4 |
| 30–34 | 2 | 1.0 |
| 35+ | 0 | 0.5 |

[a] From Clemmesen and Hjalgrim-Jensen (1981b)

compared to only 4.7 expected, were observed 20 or more years after admission for epilepsy. However, the authors quote other series in which brain tumours occurred after equally long prodromal epileptic symptoms. Certainly, the great majority of the patients with brain tumours in this series appear to have been selected into the series by virtue of their prodromal epileptic symptoms.

(2) The substantial deficit of cancers after 10 years of treatment reported in the 1974 paper has now largely disappeared. After exclusion of tumours of the central nervous system, a total of 396 cancers was observed and 406.1 expected; 10 years or more after admission, 306 were observed and 327.0 expected.

(3) In the five-year interval since the previous follow-up, two additional cases of angiosarcoma of the liver were identified in patients who had received thorium dioxide, bringing the total of such cases in the series to 10. The intervals between angiography and liver cancer ranged between 14 and 43 years. There were three cases of primary liver cancer with no history or evidence of thorotrast administration; the expected number was 3.4.

(4) There was a modest excess of cancer of the respiratory tract, which was present in patients of both sexes but not significant in either. When the results of both sexes were combined, with adjustment for age and sex, the association was significant ($\chi^2 = 5.0$), and the point estimate of the rate ratio was 1.3.

A somewhat similar study was carried out by White et al. (1979), based on 2099 epileptic patients admitted to the Chalfort Center for Epilepsy in Buckinghamshire, UK, who took anticonvulsants between 1931 and 1971. The main anticonvulsant drugs used were phenobarbital, phenytoin and primidone. Follow-up was conducted through the National Health Service Central Register and the Office of Population Censuses and Surveys. Mortality was assessed between 1 January 1951 and 31 December 1977. A total of 78 cancer deaths were identified, compared with an expected number of 50.1. Even after exclusion of the six cases of central nervous system tumours, the excess for all cancers is statistically significant, with a point estimate of the rate ratio of 1.4 and a 95% confidence interval of 1.1–1.8. The only individual cancer that occurred in significant excess, other than central nervous system tumours, was breast cancer, with eight cases observed, 3.6 expected, and a rate ratio of 2.2 (1.0–4.4). For eight of the nine other sites examined, the rate ratios exceeded 1.0, but for no other did it do so significantly. There was no case of primary liver cancer in this series.

Other studies of the relationship between use of phenobarbital and cancer risk have been of different design. Gold *et al.* (1978) evaluated the role of barbiturates in the etiology of brain tumours in children by interviewing the mothers of cases diagnosed in Baltimore, USA, between 1965 and 1975 and of two groups of control children. One control group consisted of children with no known malignant disease, who were selected on the basis of birth certificates and matched to the children with brain tumours on state of residence, race, sex and date of birth. The second control group consisted of children with malignant diseases other than brain tumours; they were matched to the children with brain tumours on the same variables, but their source or sources are not described. A total of 127 eligible brain tumour patients were identified, and 73 matched pairs of brain tumour patients and normal controls and 78 pairs of brain tumour cases and cancer controls were formed. There are additional losses in most analyses because of missing information in one member of some pairs. Positive associations with both maternal barbiturate use during the relevant pregnancy and with barbiturate use by the affected child are reported.

The associations with use of barbiturates by the affected child are not statistically significant, either when the comparison is with the normal controls or when it is with the cancer control. In both instances it rests on five discordant pairs favouring the cases and two the controls. Even if it were significant, the argument of Clemmesen and Hjalgrim-Jensen (1981b) that it reflects the occurrence of epileptic symptoms as prodromata to the cancer is compelling. Although Gold *et al.* state that 'it appears unlikely that the children were using barbiturates prior to diagnosis for symptoms which were due to the tumor', this is reasonably convincing only for one child, who had been given barbiturates 'for concussion' 13 years prior to diagnosis of the brain tumour. In all the other cases described, the possibility could not be excluded that the symptoms presaged the presence of the tumour, as the authors recognize.

Such an explanation does not apply to the reported excess use of barbiturates by the mothers during pregnancy. This association was not significant when the normal controls were used for comparison: there were six discordant pairs, four favouring the cases and two the controls. However, in comparison with the cancer controls (71 matched pairs), there were six discordant pairs, in all of which the case was affected ($p < 0.03$). Heinonen *et al.* (1977) reported on the frequency of childhood tumours (all types) among the offspring of 2413 women who took barbiturates during the first four lunar months of pregnancy; seven tumours were observed and 11.0 expected. Among the children of the 1415 women who took phenobarbital as the barbiturate, seven tumours were observed and 6.7 expected. The finding of Gold *et al.* with respect to barbiturate use in pregnancy needs to be confirmed by other data before it can be accepted.

The fourth body of published data on this matter comes from what is frankly represented as a 'hypothesis-generating study' (Friedman, 1981). By computer linkage of pharmacy and other medical care records in the San Francisco, USA, medical offices of the Kaiser-Permanente Medical Care Program, a cohort of 143 574 patients

Table 2. Standardized mortality rates (SMR) for selected cancers associated with use of three barbiturates[a]

| Cancer site | Pentobarbital sodium | | | Phenobarbital | | | Secobarbital sodium | | |
|---|---|---|---|---|---|---|---|---|---|
| | Obs. | Exp. | SMR | Obs. | Exp. | SMR | Obs. | Exp. | SMR |
| Lung | 34 | 12.3 | 2.8*** | 44 | 28.9 | 1.5 | 27 | 15.0 | 1.8** |
| Ovary | 0 | 1.8 | 0.0 | 8 | 4.5 | 1.8 | 6 | 2.1 | 2.9* |
| Thyroid | 4 | 0.9 | 4.6* | 4 | 2.1 | 1.9 | 1 | 1.1 | 0.9 |
| Lymphosarcoma | 4 | 1.8 | 2.3 | 0 | 4.2 | 0.0* | 4 | 2.1 | 1.9 |
| Pancreas | 7 | 3.9 | 1.8 | 11 | 8.5 | 1.3 | 8 | 4.1 | 1.9 |
| All sites | 112 | 85.4 | 1.3** | 235 | 211.6 | 1.1 | 122 | 100.8 | 1.2* |

[a] From Friedman (1981)
*** $p<0.002$; ** $p<0.01$; * $p<0.05$

who had had at least one medical prescription filled at the pharmacy was followed. The first batch of 95 drugs screened was selected on the basis of having been dispensed to at least 1000 patients. Among these were three barbiturates (pentobarbital sodium, phenobarbital and secobarbital), all of which showed a statistically significant association with the subsequent occurrence of cancer of the lung. Some other sites of cancer were also elevated in users of one or another of these drugs (Table 2). In addition, there was a significant association with cancer of the pancreas when the three drugs were considered as a group. The association of barbiturates (all three drugs combined) with lung cancer was found both in women (34 observed, 15.9 expected, $p < 0.002$) and in men (53 observed, 34.3 expected, $p < 0.01$).

In a previous analysis of Kaiser-Permanente subscribers, it had been found that smokers reported barbiturate use more often than did nonsmokers (Seltzer et al., 1974). Smoking data from multiphasic examinations of the cohort members were therefore linked to those from the pharmacy cohort, with the results shown in Table 3. Friedman concluded that 'Inasmuch as excess cases of lung cancer were observed in barbiturate users among nonsmokers as well as in the other groups, these data, admittedly very limited in numbers of subjects, gave no indication that smoking habits were the principal explanation of the barbiturate-lung cancer association.' But the numbers–particularly of nonsmokers and ex-smokers–are indeed limited, and an equally acceptable interpretation of these data might be that the association (or at least its statistical significance) is entirely dependent on the group of current smokers for whom no data are presented on the possibility of confounding by amount or duration of smoking.

Charts were reviewed for the lung cancer cases who were barbiturate users and for a sample of barbiturate users who did not have lung cancer, matched for sex and age. No significant difference was found with respect to year when barbiturates were first used or the estimated duration, regularity or frequency of use of any of the three compounds, suggesting no dose-response characteristic to the relationship. The short

Table 3. Observed and expected numbers of lung cancers by smoking category. All barbiturate users[a]

| Smoking category | No. of lung cancers | | Morbidity ratio |
|---|---|---|---|
| | Observed | Expected | |
| Nonsmokers | 4 | 2.7 | 1.5 |
| Ex-smokers | 6 | 4.4 | 1.4 |
| Smokers | 32 | 19.7 | 1.6 |

[a] From Friedman (1981)

average durations of use are of interest; duration of use was less than one year for 75–84% and less than one month for 44–59% for patients for whom duration could be estimated. These two features would seem to argue against the relationship being causal.

In the absence of knowledge of how the smoking habits of their cohorts compare with those of the populations with which they are compared, the studies of Clemmesen and Hjalgrim-Jensen (1981a) and of White et al. (1979) are essentially uninformative on the matter of phenobarbital and lung cancer. Noted earlier, however, was a marginally significant excess of respiratory cancer in the data of Clemmesen and Hjalgrim-Jensen, and a similar excess is seen in the data of White et al. (rate ratio, 1.4; 95% confidence interval, 0.9–2.1).

## DISCUSSION

The case of phenobarbital illustrates one of the difficulties of interpreting 'negative' epidemiological evidence—that negative epidemiological evidence is seldom unequivocally so. On the one hand, even regular use of phenobarbital in therapeutic doses appears not to be associated with any observable increase in the tumour that is supposedly induced by it in rodents. Similarly, a reasonably strong case can be made that there is no important increase (or decrease) in cancers overall in humans who are long-term users of the drug. On the other hand, until it has been resolved whether or not the association of lung cancer with barbiturate use is the result of residual confounding by amount or duration of smoking, this apparent relationship must be regarded with suspicion, in spite of the lack of dose-response relationship and the short durations of use reported in cases.

## REFERENCES

Clemmesen, J., Fuglsang-Frederiksen, V. & Plum, C.M. (1974) Are anticonvulsants oncogenic? *Lancet, i,* 705–707

Clemmesen, J. & Hjalgrim-Jensen, S. (1977) *On the absence of carcinogenicity to man of phenobarbital.* In: Clemmesen, J., ed., *Statistical Studies in the Aetiology of*

*Malignant Neoplasms, V. Trends and Risks, Denmark 1943-72,* Copenhagen, Munksgaard, pp. 38-49

Clemmesen, J. & Hjalgrim-Jensen, S. (1978) Is phenobarbital carcinogenic? A follow-up of 8078 epileptics. *Ecotoxicol. environ. Saf., 1,* 457-470

Clemmesen, J. & Hjalgrim-Jensen, S. (1980) *Epidemiological studies of medically used drugs.* In: Clemmesen, J., Conning, D.M., Henschler, D. & Oesch, F., eds. *Quantitative Aspects of Risk Assessment in Chemical Carcinogenesis,* Berlin, Springer-Verlag

Clemmesen, J. & Hjalgrim-Jensen, S. (1981a) Does phenobarbital cause intracranial tumors? A follow-up through 35 years. *Ecotoxicol. environ. Saf., 1,* 255-260

Clemmesen, J. & Hjalgrim-Jensen, S. (1981b) Brain tumors in children exposed to barbiturates. (Letter to the editor) *J. natl Cancer Inst., 66,* 215

Friedman, G.D. (1981) Barbiturates and lung cancer in humans. *J. natl Cancer Inst., 67,* 291-295

Gold, E., Gordis, L., Tonascia, J. & Szklo, M. (1978) Increased risk of brain tumors in children exposed to barbiturates. *J. natl Cancer Inst., 61,* 1031-1034

Heinonen, O.P., Slone, D. & Shapiro, S. (1977) *Birth Defects and Drugs in Pregnancy,* Littleton, MA, PSG Publishing Co.

Seltzer, C.C., Friedman, G.D. & Siegelaub, A.B. (1974) Smoking and drug comsumption. *Am. J. public Health, 64,* 466-473

White, S.J., McLean, A.E.M. & Howland, C. (1979) Anticonvulsant drugs and cancer. A cohort study in patients with severe epilepsy. *Lancet, ii,* 458-461

# PHENOBARBITAL:

## CONCLUSION

Rapporteur: R. SARACCI

The first discussion was on whether phenobarbital could induce enzymes to metabolize carcinogens in man: there was evidence for this in animals, but the situation in man was unclear. The discussion moved to brain cancer, and it was suggested that a possible explanation for the excess risk of brain cancer that had been associated with the use of phenobarbital in some studies was that epileptic groups include some patients with brain tumours—which explains both the epileptic fit and the use of drugs to control them. In reply to a number of specific questions on this aspect, Professor MacMahon expressed surprise that White *et al.* and Friedman had found only a small proportion of cases of brain tumours in their series, while Clemmesen in Denmark had found a higher proportion. Perhaps the Danish patients had been admitted and treated at a very early stage after the first epileptic fit and the group therefore included cases of brain tumour that would have been excluded from the other two studies.

Dr Jensen pointed out that during an examination of records on brain tumours in the Danish Cancer Registry, quite a number were found that would not normally be classified as neoplasia, such as, for example, vascular malformations. This could provide an additional explanation for the apparent link between 'brain tumours' and use of phenobarbital.

Another possibility that was raised was that people who had epilepsy were bound to be X-rayed frequently in the search for intracranial lesions, and that the exposure to X-rays might be the cause of the increased risk for brain tumour. In support of this idea, it was noted that Sheila Darby had just carried out a detailed analysis of the Hiroshima/Nagasaki data and had found that, after the bone marrow, the central nervous system (cord and nerves) was the most susceptible target for tumour induction.

An interesting observation from Western Australia was that it was the practice of the mental health service to give institutionalized patients a weekly ration of cigarettes, regardless of whether or not the patients smoked. Distribution of free cigarettes in this way could, of course, lead to a relatively high prevalence of smoking. If the practice were also current at the Danish and UK institutions involved in the phenobarbital studies, it is possible that the excess risk of lung cancer could have been due to the increased extent of smoking.

In summary, it was felt that the estimates of increased relative risk of cancer observed in the study of White *et al.* should not be interpreted as being causally related to exposure to phenobarbital. The increased risks were small and, in general, lack biological plausibility—particularly in the light of the fact that no excess of primary liver cancer has been reported and that this was the only organ affected in animal carcinogenesis experiments.

As to brain tumours, several factors (particularly the fact that phenobarbital is used for treatment of epileptic fits resulting from the presence of a tumour) could have accounted for the observed associations. The isolated findings in a subset of data from one case-control study, reporting an excess use of barbiturates during pregnancy by mothers of children with brain tumours as compared to women giving birth to children with other tumours, must be confirmed by other studies. As to lung cancer, the modest excess seen in three studies could be accounted for by uncontrolled confounding due to smoking. In one of the studies, however, the excess was present in all smoking categories, including non-smokers, and this needs further investigation.

The group concluded that phenobarbital should be classified among the compounds for which, at present, the epidemiological evidence weighs so strongly against the possibility that it is carcinogenic to man that one can disregard, for practical purposes, laboratory evidence that is of doubtful generality. This conclusion was, however, qualified by the unresolved and unexpected findings relating to lung cancer.

# ISONIAZID

# ISONIAZID:

# LABORATORY EVIDENCE

## P. SHUBIK
*Green College, Oxford, UK*

The first demonstration of the potential carcinogenicity of isoniazid (isonicotinic acid hydrazide) was made by Hungarian investigators (Juhász *et al.*, 1957), who reported that oral administration of isoniazid could enhance the incidence of lung adenomas and lymphomas in mice. Numerous efforts have been made to repeat this study; it has been found consistently that the incidence of lung adenomas is enhanced, but the enhancement of lymphomas has not been confirmed. In one study by Toth and myself in 1966, it was found that in C3H mice with a high mammary cancer incidence isoniazid could still give rise to lung adenomas (a rare tumour in this strain) but virtually totally inhibited the occurrence of mammary tumours.

In other species isoniazid does not seem to be active: studies in hamsters gave negative results; one study in rats showed the occurrence of liver, mammary and lung tumours, however, the incidences in the liver and lung were not significant.

Interpretation of these results is complicated by the results of studies with hydrazine sulfate, which is the principal metabolite of isoniazid. Hydrazine sulfate gave rise to a significant incidence of lung and liver tumours—many more than with isoniazid—but not of mammary tumours. Other studies in rats have had negative results.

The only clear-cut effect of isoniazid in respect of carcinogenesis appears to be the induction and enhancement of lung adenomas. These tumours are extremely common in mice, and some strains have a 100% incidence. They have no counterpart in human pathology. In the mouse, they frequently progress to carcinoma, but there have been major divisions of opinion among pathologists as to their cell of origin and to criteria for determining malignancy. All efforts to find an etiology for these tumours in mice have failed so far—common sense would lead one to believe that a viral origin was likely, but this appears not to be the case. Certain workers in carcinogenesis have recommended the use of these tumours as a screening tool in the investigation of potential carcinogens, since there is certainly a high level of correlation between the enhancement of lung adenomas and the induction of other tumours.

One need only be reminded of the case of urethane (ethyl carbonate), which for many years was thought to have a very restricted activity—only enhancing the incidence of lung adenomas in certain strains of mice. After many years it was found that prolonged application revealed a potential for inducing many tumours in many species. Perhaps the current restricted profile of isoniazid as a compound that can only 'induce' lung adenomas would change with further investigations. These seem unlikely to be forthcoming.

It is of interest to recognize that subsequent to the discovery of the 'carcinogenicity' of isoniazid a variety of hydrazide derivatives have been investigated and some found to be carcinogenic.

There is no doubt that at present no drug company would develop a drug that had these adverse effects. Considering the benefits derived from isoniazid so far, consideration must be given to the appropriateness of our present approach.

## REFERENCES

Juhász, J., Baló, J. & Kendrey, G. (1957) Über die geschwulsterzeugende Wirkung des Isonicotinsäurehydrazid (INH). *Z. Krebsforsch.,* **62,** 188–196

Toth, B. & Shubik, P. (1966) Carcinogenesis in Swiss mice by isonicotinic acid hydrazide. *Cancer Res.,* **26,** 1473–1475

# ISONIAZID:

# EPIDEMIOLOGICAL EVIDENCE

## T.W. ANDERSON

*Department of Health Care and Epidemiology, University of British Columbia, Vancouver, BC, Canada*

Isoniazid was introduced to clinical practice in 1952, and the possible carcinogenic effects of this drug have been examined in approximately 14 published studies that appeared between 1956 and 1980 (Table 1).

It is a truism that it is impossible to prove a negative, but the larger the study the more confident one can be that if there is an undesirable side-effect, its magnitude must be extremely small. An initial 'triage' of the published studies listed in Table 1 shows that among those with cancer mortality as the end-point, seven have been based on 10 cases or less, and, of the remaining seven, three (Ferebee no. 4, Glassroth no. 7 and Costello no. 11) are based on essentially the same study population. The number of substantial studies is therefore quite small, but isoniazid is unusual among potential carcinogens in that data are available from large-scale, randomized, placebo-controlled trials—the strongest type of epidemiological evidence.

In the penultimate column of Table 1 are listed the approximate ranges of follow-up periods from the time the drug was first administered to the closing date of the death certificate (or other) search. It should be noted that, although several of the published studies have maximal follow-up periods of over 20 years, the minimal follow-up period is often well below this figure.

In most of the reports, mortality has been the only end-point studied, but morbidity was examined in two (Howe no. 9, Costello no. 11). Similarly, while cancer mortality has been the major focus of attention, two studies (Howe no. 9, Boice no. 10) have also reported on some possible non-cancer long-term side-effects.

The dosage of isoniazid appears to have been fairly uniform, at approximately 4–5 mg/kg body weight per day.

For convenience, the published studies are discussed under the following headings:
    Small-scale observational (both cohort and case-control)
    Large-scale observational
    Randomized placebo-controlled

Table 1. Published cohort and case-control studies of cancer incidence in human populations exposed to isoniazid therapy, by year of publication[a]

| Reference number | Year published | First author | Comments | Follow-up (no. of years) | Number of cancers in those exposed to isoniazid |
|---|---|---|---|---|---|
| **Cohort** | | | | | |
| 1 | 1956 | Pompe | | 1–4 | 7 |
| 2 | 1967 | Hammond | | 1–14 | 10 |
| 3 | 1968 | Nyfors | | 11–15 | 8 |
| 4 | 1969 | Ferebee | Randomized placebo-controlled | 7–9 | 349 |
| 5 | 1970 | Campbell | Lung cancer only | 9–14 | 17 |
| 6 | 1976 | Stott | | 16–21 | 98 |
| 7 | 1977 | Glassroth | Update of no. 5 | 8–14 | 404 |
| 8 | 1979 | Clemmesen | | 14–24 | 253 |
| 9 | 1979 | Howe | Mortality | 13–21 | 1135 |
|   |      |      | Morbidity | 9–21 | 535 |
| 10 | 1980 | Boice | | 8–23 | 8 |
| 11 | 1980 | Costello | Randomized placebo-controlled (subset of no. 5; morbidity only) | 16–19 | 74 |
| **Case-control** | | | | | |
| 12 | 1977 | Glassroth | Bladder & renal cancer | 1–22 | 0 |
| 13 | 1978 | Miller | Bladder cancer | 1–23 | 8 |
| 14 | 1979 | Sanders | All sites (children) | 1–15 | 7 |

[a] All based on mortality only, except for those by Howe et al. (no. 9), in which both mortality and morbidity were examined, and Costello et al. (no. 11), in which only morbidity was analysed.

## SMALL-SCALE OBSERVATIONAL STUDIES

*Cohort studies* (Reference nos 1, 2, 3, 5, 10, Table 1)

The first report of a study in humans linking isoniazid to cancer was that of Pompe, published in 1956, barely four years after introduction of the new drug. This report concerned a group of 150 patients with lupus vulgaris, of whom seven had developed cancerous changes in their lupoid lesions, compared with only one or two predicted on the basis of previous experience. However, these patients had also received a variety of other treatments, including radiotherapy; and in view of the almost uniformly negative findings of later studies, together with the very short latent period, the results of this study must be accepted with some caution.

Subsequently, Nyfors (1968) (no. 3) followed up until 1967 some 245 patients with lupus who had been treated with isoniazid between 1952 and 1956. A total of 51 deaths was observed, of which eight had been ascribed to cancer. On the basis of age- and sex-specific rates for the Danish population, six cancer deaths would have been expected in this group, i.e., a slight but nonsignificant excess was present in the isoniazid group.

In 1967, Hammond et al. (no. 2) reported three investigations of the isoniazid/cancer question in a single paper. The first study involved a follow-up of the approximately 19 000 individuals who had recorded a past history of tuberculosis in the massive (more than one million) study of the effects of smoking conducted by the American Cancer Society in 1959-1960, and for which there had been a 98% follow-up to September 1964. It was estimated that 5-10% of those with a history of tuberculosis would have received isoniazid, but no individual record of such history was available. A total of 268 cancer deaths was recorded for the tuberculosis group, compared with an age- and sex-adjusted figure of 262.1 expected on the basis of the mortality experience of the nontuberculous subjects. Since these data relate more to the question of whether tuberculous patients experience an elevated death rate from cancer, irrespective of treatment with isoniazid or other drugs, these results contribute only indirectly to the present discussion and are not pursued.

The second part of the paper by Hammond et al. reported the experience of 311 tuberculosis patients in the private practices of two of the authors. Treatment with isoniazid was begun between October 1951 and September 1956, and mortality follow-up was to June 1966. A total of 10 patients had died of cancer, compared with an expected figure of 6.3 based on appropriate age- and sex-adjusted rates for the states of New York and New Jersey, USA. In addition to the excess of observed over expected, there was also something of a latency gradient, with an observed : expected ratio of 4 : 3.1 for the first seven years and 6 : 3.1 for the later years. However, there was no evidence of a dose-related effect, since half of the cancer cases had been treated for less than two years, and half for more than two years (at a standard dose of 4 mg/kg per day).

The third piece of evidence, submitted by Hammond et al., concerned 502 women who had received isoniazid during pregnancy, between the years 1953 and 1966. By 1 June 1966, none of the women or their offspring had developed cancer, compared with an expected incidence of about one case.

The paper in 1970 by Campbell and Guilfoyle (no. 5) was largely a follow-up of a previous observation by Campbell of an elevated cancer death rate among tuberculosis patients (Campbell, 1961). The study group was made up of 3064 Australian ex-servicemen who had at some time been diagnosed as having pulmonary tuberculosis, and who were alive and resident in the state of Victoria on 1 January 1961. Cause of death was then obtained for all those who died during the subsequent five years. Expected deaths were calculated on the basis of age-specific general population death rates. A 10% sample of the initial cohort was selected, and records examined to determine the frequency of smoking and exposure to isoniazid. Most of the paper is concerned with observed deaths in the total group of tuberculosis patients, which are compared with expected figures based on general population death rates in the state of Victoria; only a sample estimate was available of the proportion treated with isoniazid. It was concluded that of the 17 cases of lung cancer observed, 15.5 could reasonably be attributed to cigarette smoking, leaving a small and statistically non-significant excess of 1.5 cases that might have been related to isoniazid.

In 1980, a study published by Boice and Fraumeni (no. 10) used data collected originally for a study of the possible carcinogenic effects of ionizing radiation (from pneumothorax treatments) in female patients admitted to sanatoria in Massachussets, USA, between 1930 and 1954. Because isoniazid was not introduced until 1952, only the 1482 women known still to be alive in that year were included in the study cohort. By 1975, it was found that 17% had died and 3% could not be traced. Among 338 women known to have received isoniazid treatment, 63 had died, eight of cancer. The number of cancer deaths expected on the basis of general Massachussets mortality rates was 8.8. None of the eight observed cancer deaths was due to lung or bladder cancer.

A parallel finding of some interest was of five deaths from cirrhosis of the liver, compared to 0.8 expected. Although based on small numbers, the excess showed evidence of a gradient with both total dose (duration of treatment) and latent interval. Thus, four of the five patients had been treated for over one year, compared to 0.3 to be expected from general treatment patterns; and the ratio of observed : expected increased with interval from beginning of treatment: 0–4 years, 0 : 0.2; 5–9 years, 1 : 0.2; 10–14 years, 3 : 0.2; 15+ years, 1 : 0.1. The authors point out that not only are these findings similar to those of Howe *et al.* (1979) (discussed later) but they are biologically plausible, since not only is frank hepatitis recognized as an occasional acute side effect of isoniazid therapy, but enzymic evidence of disturbed hepatic function can be shown in 10–20% of patients who received isoniazid.

Finally, mention should be made of two other small-scale cohort studies that are sometimes referred to in articles on this subject but have not been published in the general scientific literature. One, presented at a conference in 1969 by Kerby *et al.*, is quoted by Glassroth *et al.* (1977a) (no. 7) as showing no evidence of isoniazid carcinogenicity in 1196 men followed for a mean of 57 months. The second study, by Kerr and Chipman, presented at a conference in 1976, was referred to by Miller *et al.* (1978) (no. 13) as having shown an excess of lung and bladder cancers. The authors have kindly given me permission to quote from their unpublished manuscript, which shows that 2470 tuberculous subjects, of whom 878 had received isoniazid, were followed up for a period of 10 to 20 years. Data from the Connecticut Tumor Registry were used to create expected numbers. Among the isoniazid-treated cases, there were 48 observed cancers, and 45.3 were expected. For lung cancer, the observed : expected ratio was 10 : 3.15, and for bladder cancer it was 2 : 1.15. The subjects who had not received isoniazid had an observed : expected ratio for all cancers of 55 : 94.9, for lung 3 : 3.9, and for bladder 6 : 8.73. The authors also noted that male cases of genitourinary cancer had, on average, received isoniazid for a longer period of time than the average for the group.

*Case-control studies* (Reference nos 12, 13, 14, Table 1)

In the course of a case-control study of bladder cancer, Miller *et al.* (1978) (no. 13) looked at a number of possible etiological agents, including isoniazid. A preliminary report of this study prompted Glassroth *et al.* (1977b) (no. 12) to carry out a case-control study of bladder and renal cancers, which was finally published some two years earlier than the final report of Miller *et al.* (1978).

The Glassroth *et al.* (1977b) study identified 142 cases of bladder cancer and 48 of renal cancer that had occurred by 1974 in a population of approximately 90 000 persons who had been the subject of a special census in Weston, Maryland, USA, in 1963. For each case, two controls of the same age, race and sex were selected randomly from census lists. The tuberculosis treatment records of the county were then searched for evidence of isoniazid administration. Of the total 570 cases and controls, only two individuals—both controls—were found to have ever received isoniazid.

Miller *et al.* (1978) collected information on 256 cases of bladder cancer in the Ottawa, Canada, area and on 512 controls, chosen from among other urological patients and (for the over 75-year-age group) from other ambulatory clinics in the same area. Eight cases had a history of exposure to isoniazid compared to 14 controls: 5 *versus* 12 males and 3 *versus* 2 females.

As a matter of interest, if the figures from the Miller *et al.* (1978) and Glassroth (1977b) studies are combined, there were eight cases and 16 controls with exposure to isoniazid, for an overall relative risk of exactly 1.0.

The third case-control study (Sanders & Draper, 1979) (no. 14) was not truly 'small-scale', in that the authors used data from the Oxford Survey of Childhood Cancers, in which there were 11 169 matched case-control pairs of children. An earlier study of these data had shown a high ratio of case mothers with a history of tuberculosis (27 *versus* 13 control mothers), and the possibility that this excess was related to isoniazid treatment was explored by examining medical records. Unfortunately, records could be found for only 17 of the 27 tuberculous mothers and 9 of the 13 control mothers, but the proportions that had received isoniazid during their pregnancy were quite similar—41% (7 out of 17 cases) and 56% (5 out of 9 controls).

*Summary*

These small-scale studies showed a mixture of negative and (usually non-significant) positive results. This pattern is consistent with either a weakly causal relationship, or random variation around a truly nil effect.

## LARGE-SCALE OBSERVATIONAL STUDIES

The first reasonably substantial follow-up trial was that published by Stott *et al.* in 1976 (no. 6). The risk of death in patients treated with isoniazid was compared with that of the general population of England and Wales, and a parallel comparison was carried out of patients treated for tuberculosis around the same time, but in whom isoniazid had not been used.

A total of 3842 patients who had begun treatment at either of two sanatoria between 1950 and 1957 were traced to the end of 1973—a potential range of 16–21 years since isoniazid treatment. The original plan had been to compare mortality in patients first treated before the introduction of isoniazid with mortality in those treated later. However, the use of isoniazid became so widespread so rapidly that a substantial proportion of the early-treatment patients subsequently received isoniazid, thus confounding the comparison. The patterns of mortality were substantially different before and after 1952, irrespective of whether isoniazid had been used. This made interpretation of the results difficult. Overall, there had been 98 cancer deaths among the 2696 patients who had at some time received isoniazid, compared to an expected figure of 76.2, giving a standardized mortality ratio (SMR) of 129. In contrast, among the 1146 patients who had not received isoniazid at any time, there were 19 cancer deaths compared to 29.6 expected, for a SMR of 64, i.e., about half that of the isoniazid group.[1]

A bias in selection for isoniazid treatment may well have been responsible for this difference in SMRs, since increasing duration of follow-up showed no positive trend for cancer deaths: the SMR in the isoniazid group was 210 for years 1–4, 130 for years 5–8, 90 for years 9–12, 120 for years 13–16 and 140 for 16 plus years.

The total dose of isoniazid also showed no trend in the risk of cancer death, with SMRs of 150, 150, 100 and 130 for total doses of up to 50 g, 50–99 g, 100–199 g and 200 g or more, respectively.

The difference in SMRs in deaths from lung cancer was even more striking. This site accounted for 19 of the 98 cancer deaths in the isoniazid group, and 3 of 19 in the non-isoniazid group; the corresponding expected figures were 26.0 and 9.6, giving SMRs of 142 and 31, respectively. Detailed figures were not given for any other site.

In 1979, Clemmesen and Halgrim-Jensen (no. 8) published the results of a follow-up study of 3371 patients treated with isoniazid between 1950 and 1962, who had been followed until 1976. The authors ignored deaths that had occurred within the first two years after onset of isoniazid therapy, and reported a total of 363 cancer deaths, compared to 228.34 expected on the basis of general Copenhagen death rates (SMR, 159). Among the 1573 patients who had not received isoniazid, there had been 104 cancer deaths, compared to 97.45 expected, for a SMR of 107.

As with the study by Stott *et al.* (1976), duration of follow-up did not appear to influence the risk of death from cancer: from 2–4 years, observed deaths in the isoniazid group were 51 with 32.60 expected (SMR, 156); for 5–9 years, 106 : 59.42 (178); for 10–14 years, 119 : 63.44 (146); for 15–19 years, 57 : 49.33 (116); and for 20–24 years, 30 : 22.20 (135).

For lung cancer, SMRs were 348 (102 : 29.3) for the isoniazid group and 213 (19 : 8.9) for the non-isoniazid group.

The study that involved the largest number of deaths among subjects exposed to isoniazid was that by Howe *et al.* (1979) (no. 9). The Canadian National Tuberculosis Registry was used to identify admissions to tuberculosis hospitals or sanatoria for the years 1951–1960. Initially, 108 456 records were identified, but these were reduced to 64 037 after internal linkage to identify repeat admissions of the same person. The tuberculosis records were then computer-linked to the Canadian Mortality Data Base, covering the years 1952 to 1973, using, as identifying information, surname, first and second given names, day, month and year of birth, sex, and province of residence. A system of weighting factors was used to determine whether or not a true 'match' had been made. A similar linkage procedure was carried out with the Canadian Cancer Incidence Registry for the years 1969–1973. These data did not include the

---

[1] Note: Some authors use the terms relative risk (RR) and standardized mortality ratio (SMR) interchangeably. For the purpose of this review, the former is used when comparison is with a non-exposed cohort, and SMR when the comparison group is the general population.

Province of Ontario (about one quarter of the Canadian population), because Ontario incidence data were not available for these years.

The authors identified a total of 1745 treated cases of tuberculosis who had died of cancer by 1973. Of these, 1135 had a record of isoniazid treatment, while 610 had not been exposed to isoniazid. (However, it must be noted that some of the latter may have been exposed to isoniazid after 1960, so that the distinction between exposed and unexposed may be somewhat blurred. In addition, some patients may have received isoniazid on an outpatient basis, and this exposure may not have been recorded.)

Standardized mortality and incidence ratios were calculated for various causes, by sex, age and calendar period, using the exposed individuals as the standard population distribution. In passing, it may be noted that elsewhere in their paper the authors used a *proportionate* mortality approach to compare overall mortality rates in the tuberculosis patients with rates in the general Canadian population, because over- or undermatching leading to over- or underestimates of death rates may have occurred, this being an inherent problem in linkage programmes. The degree of over- or undermatching will depend on the level at which the weighting cut-off is set for a 'true match'. However, internal comparisons, such as isoniazid therapy *versus* no isoniazid therapy should not be affected (since bias would probably be the same in each group), so direct comparison of rates should be acceptable.

When patients exposed to isoniazid were compared with nonexposed, there were 1135 cancer deaths compared to approximately 1149 expected, yielding a RR of 0.99. No individual cancer site showed a convincing difference in incidence between isoniazid-treated and control subjects. Similarly, cancer incidence figures showed no evidence of an effect of isoniazid. (Note that an indeterminate number of individuals would have appeared in both the incidence and mortality figures.)

Among noncancer causes of death, it is of interest that chronic bronchitis and emphysema rates were elevated, possibly indicating that isoniazid tended to be used more readily in those patients with more severe respiratory problems. Also, the number of cirrhosis deaths was elevated among females (RR, 3.02) but not in males. (See also Boice & Fraumeni, 1980; no. 10, Table 1.)

*Summary*

The three large-scale observational studies have given somewhat inconsistent results. The largest (Howe *et al.*, 1979), showed no elevated cancer rate in isoniazid-treated patients, but the studies by Stott *et al.* (1976) and Clemmesen and Hjalgrim-Jensen (1979) showed elevated cancer rates in isoniazid-treated patients compared to nontreated individuals. It must be emphasized that the allocation to isoniazid treatment or nontreatment was not a random process in these observational studies, and it is possible that some associated abnormality accounts for the elevated cancer rates in those patients selected for isoniazid treatment compared with those who were not.

## RANDOMIZED PLACEBO-CONTROLLED TRIALS

The most scientifically satisfying and potentially 'conclusive' studies are the experimental studies of Ferebee (1969) (no. 4, Table 1), Glassroth et al. (1977a) (no. 7) and Costello and Snider (1980) (no. 11). These studies were originally designed by the US Public Health Service to test the efficacy of the prophylactic use of isoniazid in persons exposed to tuberculosis infection, and involved random allocation of large numbers of symptomless individuals (some, but not all, of whom were tuberculin-positive) to either treatment (isoniazid) or control (placebo) groups.

Two main groups were studied for possible long-term side-effects of isoniazid: 25 033 household contacts of newly diagnosed cases of tuberculosis, and 27 924 patients in mental institutions. For each group, allocation to isoniazid or placebo was made not on the basis of individual randomization, but—for administrative simplicity and reliability—households or (for mental patients) wards.

The authors do not comment on the degree to which the studies were double-blind (e.g., whether the placebo was a convincing imitation of the active drug) or the degree of compliance in either group, but it is implied that compliance was good (or at least similar) and that patients were not aware of the nature of their medication.

Enrolment was from 1957 to 1959 for household contacts, and 1957 to 1960 for mental patients. If a mental patient changed wards, and thus the nature of the treatment, he or she was dropped from the follow-up study. This reduced the total number of mental patients from 27 924 to 25 210.

In the first follow-up report (Ferebee, 1969) (no. 4), deaths were identified up to December 1967. Cancer deaths in isoniazid *versus* placebo groups were 70 *versus* 66 in the contact group, and 279 *versus* 264 in the mental patients.

By the time of the report by Glassroth et al. (1977a) (no. 7), follow-up had been extended to December 1970 for household contacts but only to July 1968 for mental patients. Cancer deaths in isoniazid *versus* placebo groups now totalled 108 *versus* 109 for contacts and 296 *versus* 278 for mental patients. (The smaller number of deaths in the contact group was largely a reflection of age, many being children or young adults.) For this discussion only the later figures (Glassroth *et al.*, 1977a) are considered, since the Ferebee data are contained within them.

Due in part to the randomization by households rather than individuals, treatment and control groups of the household contacts were not as similar in some respects as might have been expected with such large numbers, so that adjustment for age and sex produced a reversal of the apparent cancer balance—crude rates were 8.68 and 8.65 per 1000 for isoniazid and placebo groups, respectively, while adjusted rates were 8.66 and 8.68. (Needless to say, neither of these differences approaches statistical significance.) There was no indication of a different pattern for individual cancer sites, and there was also no suggestion of an increased relative frequency in the isoniazid group with increasing latent period.

The study of mental patients involved allocation by wards, and these were later subdivided into 'intensive care' and 'normal care'. In the normal care wards, allocation to isoniazid or placebo was reasonably well-balanced, with 10 531 and 10 418 patients, respectively, but intensive care patients were weighted towards isoniazid treatment (2353) rather than placebo (1903). The fact that the overall death rate tended to be higher among the intensive care patients tended to exaggerate the number of deaths linked to isoniazid. In fact, there were 296 cancer deaths in the isoniazid group and 278 in the placebo group, with age- and sex-adjusted rates of 23.2 and 22.3 per 1000, respectively.

From the data provided by the authors it is possible to adjust for this imbalance in type of care; using the placebo group as the standard distribution, the rate in the isoniazid group is reduced to 22.6 per 1000, almost identical to the placebo rate of 22.3.

Deaths from cancer at individual sites were similar in number for both groups. Primary lung cancers accounted for 15 deaths in each group, but lung cancers 'unspecified' as to primary or secondary totalled 27 in the isoniazid group and 17 in the placebo group. More detailed follow-up would seem to be indicated to clarify this difference.

In terms of latent interval the picture is also somewhat unclear, since during the first 10 years there is a slight suggestion of an upward trend in the ratio of deaths in the isoniazid: placebo groups with increasing latent period. Thus, for up to two years of follow-up there were 80 cancer deaths in the isoniazid group compared with 81 in the placebo group (RR, 0.99); for 3–4 years, 84 : 84 (1.00); for 5–6 years, 86 : 81 (1.06); for 7–8 years, 82 : 72 (1.14); for 9–10 years, 49 : 43 (1.14); for 11–12 years, 16 : 20 (0.80); for 13–14 years, 7 : 6 (1.17).

This suggestion of an upward trend may well be an artefact, related perhaps to the fact that the intensive care group was weighted towards isoniazid treatment *and* generally higher mortality (the biennial figures quoted are *crude*); it would be highly desirable for the follow-up of this important group of subjects to be extended beyond 1968, and for an attempt to be made to more thorough adjustment for confounding variables.

Costello and Snider (1980) (#11) reported a third study of long-term effects in the same US Public Health Service trials. Their study involved cancer incidence in the Puerto Rican component of the 'household contact' trial of prophylactic isoniazid. Some 19 communities in Puerto Rico contributed 11 894 of the 25 033 subjects analysed by Ferebee (1969) and by Glassroth *et al.* (1977a), and a computer listing of these individuals was linked manually to the central Puerto Rican Cancer Registry from 1957 up to 1976. By 1976, 98% of the subjects had been in the trial for at least 17 years. Among those who had received isoniazid (initially, 5924 persons) 74 cancers were identified (excluding skin cancers), compared to 84 in the placebo group (5970 persons). Thus, although the difference was not statistically significant, the incidence of cancer was actually lower in the isoniazid group. Furthermore, there was no evidence of a relative increase in the frequency of cancers in the isoniazid group with increasing duration of observation. Cancer incidence rates were approximately the same up to year 12, with a lower rate for the isoniazid group in subjects followed for between 14 and 19 years. By 1976, there were 49 fatal cancers in each group (some of which would have been included in the previous reports).

*Summary*

These experimental trials largely avoid the problems of a possible association between the presence of active tuberculosis and subsequent development of cancer. They provide the 'cleanest' epidemiological evidence of an absence of carcinogenicity following one year's exposure to isoniazid. It would be highly desirable to extend the follow-up period of observation, partly because it is possible that isoniazid might manifest a carcinogenic effect only after more than 20 years, and partly because there is a slight (but quite possibly artefactual) suggestion of a latency gradient in the early mortality figures.

## CONCLUSION

Of the 14 published studies that have assessed the possible carcinogenicity of isoniazid, three have been based on a large population assigned randomly to active drug or placebo. Their main drawback is that although maximum follow-up is now potentially more than 25 years, the most recently published data involve a maximal follow-up of only 14 years (mortality) or 19 years (morbidity). Otherwise, this group of studies provides the most convincing evidence to date that isoniazid is not carcinogenic.

There have also been three large-scale observational studies. The largest of these showed no evidence of carcinogenicity, but in the other two the isoniazid-treated patients experienced more cancers than those not so treated. The absence of any gradient for duration or dose suggests that the therapeutic selection process may have led to this imbalance.

The remaining published studies have been of too small a scale to provide useful evidence other than to exclude a gross carcinogenic effect or a long latent interval in carcinogenesis. It is important that follow-up of exposed patients be continued, particularly of those enrolled in the experimental randomized trials.

In addition to possible carcinogenic effects, two studies have detected possible long-term liver toxicity, with elevation of the death rate from cirrhosis of the liver.

## REFERENCES

Boice, J.D. & Fraumeni, J.F. (1980) Late effects following isoniazid therapy. *Am. J. public Health,* **70,** 987–989

Campbell, A.H. (1961) The association of lung cancer and tuberculosis. *Austr. Ann. Med;* **10,** 129–132

Campbell, A.H. & Guilfoyle, P. (1970) Pulmonary tuberculosis, isoniazid and cancer. *Br. J. Dis. Chest,* **64,** 141–149

Clemmesen, J. & Hjalgrim-Jensen, S. (1979) Is isonicotinic acid hydrazide (INH) carcinogenic to man? *Exotoxicol. environ. Saf.,* **3,** 439–450

Costello, H.D. & Snider, D.E. (1980) The incidence of cancer among participants in a controlled, randomized isoniazid prevention therapy trial. *Am. J. Epidemiol.,* **111,** 67–74

Ferebee, S. H. (1969) Controlled chemoprophylaxis trials in tuberculosis. *Adv. Tuberc. Res., 17,* 88–106

Glassroth, J. L., White, M. C. & Snider, D. E. (1977a) An assessment of the possible association of isoniazid with human cancer deaths. *Am. Rev. resp. Dis., 116,* 1065–1074

Glassroth, J. L., Snider, D. C. & Comstock, G. W. (1977b) Urinary tract cancer and isoniazid. *Am. Rev. resp. Dis., 116,* 331–333

Hammond, E. C., Selikoff, I. J. & Robitzek, E. H. (1967) Isoniazid therapy in relation to later occurrence of cancer in adults and infants. *Br. med. J., ii,* 792–795

Howe, G. R., Lindsay, J., Coppock, E. & Miller, A. B. (1979) Isoniazid exposure in relation to cancer incidence and mortality in a cohort of tuberculosis patients. *Int. J. Epidemiol., 8,* 305–312

Miller, C. T., Neutel, C. I., Nair, R. C., Marrett, L. D., Last, J. M. & Collins, W. E. (1978) Relative importance of risk factors in bladder carcinogenesis. *J. chronic Dis., 31,* 51–56

Nyfors, A. (1968) Lupus vulgaris, isoniazid and cancer. *Scand. J. resp. Dis., 49,* 264–269

Pompe, K. (1956) Einfluss von Isonicotinehydrazid auf die Lupuskarzinomentstehung. *Dermatol. Wochenschr., 133,* 105–107

Sanders, B. M. & Draper, G. J. (1979) Childhood cancer and drugs in pregnancy. *Br. med. J., i,* 718–719

Stott, H., Peto, J., Stephens, R. (1976) An assessment of the carcinogenicity of isoniazid in patients with pulmonary tuberculosis. *Tubercle, 57,* 1–15

# ISONIAZID:

# CONCLUSION

Rapporteur: J. HIGGINSON

The question was raised whether in the studies described other drugs had been administered concurrently with isoniazid. Many authors had made no mention of other drugs; streptomycin seemed to be the only other drug of consequence. However, since isoniazid had been used prophylactically in a randomized clinical trial, it was unlikely that other drugs had had significant effects.

The higher frequency of cirrhosis reported in certain studies was reviewed briefly, and the possibility that alcohol or hepatitis are confounding factors associated with either the illness or the treatment was suggested.

There was general discussion on the problems associated with performing randomized trials and the statistical analysis of household *versus* individual data. While important, it did not appear that there was any significant effect on the final results. The definite excess of cancer in the two large studies was emphasized, but it was pointed out that there was no consistency as to the site involved, which made it difficult to draw conclusions. The advantages of randomized trials for studies of this type were described, but it was not felt worthwhile to waste resources on following up the studies. There was no evidence of an inhibitory effect of isoniazid on the incidence of breast cancer. Regret was expressed with regard to the relatively short duration of many of the studies and to the fact that relatively few data were available on long intervals. There was a consensus that further trials were unlikely to be attempted. The group concluded that the evidence was inadequate to permit a firm conclusion but suggested that isoniazid was unlikely to have produced a quantitatively large increase in risk under the conditions of exposure that have operated in the past.

# NITRATES

# NITRATES:

# LABORATORY EVIDENCE

### W.G. FLAMM

*Office of Toxicological Sciences, Bureau of Foods, Food and Drug Administration, Washington DC, USA*

## INTRODUCTION

Nitrates as such have not been shown to be carcinogenic in animals, although the adequacy of their testing has been questioned (National Academy of Sciences, 1981). Nitrates are, however, reduced by bacteria in saliva to nitrite, which has been tested extensively and in a series of studies has been alleged to induce lymphoma in rats (Newberne, 1979). Further review of the histopathology by the Universities Associated for Research and Education in Pathology, Inc. failed to confirm many of the original diagnoses of lymphoma, leading an Interagency Working Group (1980) in the USA to conclude that the available evidence was insufficient to declare nitrite a carcinogen. Nevertheless, nitrite is mutagenic to bacteria and forms inter- and intrastrand cross-links among purines in double-stranded DNA and attacks cytosine. More importantly, under acidic conditions nitrite can nitrosate amines and amides to form carcinogenic $N$-nitroso compounds. Over 300 such compounds have been tested and shown to be carcinogenic to animals under laboratory conditions. It has also been possible to produce cancer by administering nitrites and nitrosatable amines or amides at high concentrations separately, in either the diet or water.

## RESULTS

The few experiments conducted in animals have provided no evidence that nitrate is carcinogenic, although none of the studies performed to date can be considered wholly adequate by present-day standards. Nitrite has been tested numerous times, but in most cases information came from experiments in which animals being administered nitrite were used as controls in experiments designed to study the carcinogenic effects of the simultaneous administration of nitrite and an amine. The lymphoreticular effects reported by Newberne (1979) were not confirmed in an extensive review of all microslides by a joint committee of experts established by the Universities Associated for Research and Education in Pathology, Inc.

The absence of evidence that nitrite is a direct carcinogen does not diminish the importance or the possibility that it may react with dietary components or endogenous metabolites to produce carcinogenic $N$-nitroso compounds. A good deal is known about the carcinogenicity of $N$-nitroso compounds and about the reactions by which they are formed. Vitamins C and E and certain phenols inhibit most nitrosation reactions, while other phenols, thiocyanate and iodide ion catalyse nitrosation. Nitrosation kinetics depend upon the concentration of both nitrite and the nitrosatable amine or amide. Consequently, by lowering the nitrite and amine concentration by 100-fold, the level of $N$-nitroso compounds that may form is reduced one million fold in the absence of a catalyst and by ten thousand fold in the presence of a catalyst.

## SUMMARY

Neither nitrate nor nitrite *per se* has been demonstrated to produce cancer in experimental animals. The potential carcinogenicity of either of these substances would appear to derive from their involvement in the production of carcinogenic $N$-nitroso compounds. The extent to which such compounds form has been the subject of numerous studies.

## REFERENCES

Interagency Working Group (1980) *Report of the Interagency Working Group on Nitrite Research,* Washington DC, Department of Health and Human Services, Public Health Service, Food and Drug Administration

National Academy of Sciences (1981) *The Health Effects of Nitrate, Nitrite and N-Nitroso Compounds,* Washington DC, National Academy Press

Newberne, P.M. (1979) Nitrite promotes lymphoma incidence in rats. *Science,* **204,** 1079–1081

# NITRATES:

# EPIDEMIOLOGICAL EVIDENCE

## P. FRASER

*Epidemiological Monitoring Unit, London School of Hygiene and Tropical Medicine, London, UK*

There is no evidence to suggest that either nitrates, or nitrites, are carcinogenic. In contrast, of the 300 or so *N*-nitroso compounds that have been tested in experimental animals, the great majority have proved to be carcinogenic in one or more species, and often in more than one target organ. Although the value of these animal tests for predicting risk to humans is unknown, they do provide persuasive evidence that *N*-nitroso compounds are likely to be carcinogenic in man. However, as yet, they have not been definitely incriminated as the cause of any human cancer.

In the UK, current concern over rising nitrate levels in drinking-water stems principally from the fact that nitrites, which are derived mainly from bacterial reduction of ingested nitrates, may react *in vivo* with nitrosatable substrates in certain foods to form *N*-nitroso compounds. Members of one class of these substances, the *N*-nitrosamines, have been detected in very small amounts in the normal human stomach, and in larger amounts in subjects with diseases associated with low gastric acidity (Reed *et al.*, 1981). They have also been detected in infected urinary bladders (Brooks *et al.*, 1972; Hicks *et al.*, 1977; Radomski *et al.*, 1978), in saliva (Tannenbaum *et al.*, 1978) and in faeces (Wang *et al.*, 1978), although recently it has been suggested that earlier studies may have overestimated the amount synthesized in the colon (Archer *et al.*, 1982).

The amount of *N*-nitroso compounds that can be formed *in vivo* depends in part on the availability of nitrite, which is itself dependent on the availability of nitrate, the presence of a microbial population with nitrate reductase activity, and conditions favourable to chemical nitrosation (Tannenbaum, 1983). Therefore, if endogenously-formed *N*-nitroso compounds are important in human cancer, populations who ingest large amounts of nitrate might be expected to have a higher incidence of cancer of the relevant target organ. With the exception of studies of oesophageal cancer in Iran and China, and a few studies that have considered cancer risk in general, most epidemiological investigations have examined this hypothesis in relation to gastric cancer (National Academy of Sciences, 1981).

In 1979, the Royal Commission on Environmental Pollution and other independent researchers, including ourselves, looked at the results of epidemiological studies available at that time and concluded that there was then no evidence that unambiguously associated nitrates, nitrites or $N$-nitroso compounds with cancer of any organ in man (Royal Commission on Environmental Pollution, 1979; Fraser et al., 1980). The reviews took into account the results of studies relating gastric cancer risk to nitrate fertilizer usage in Chile (Zaldivar & Robinson, 1973; Armijo & Coulson, 1975; Zaldivar & Wetterstrand, 1975; Zaldivar, 1977) and to waterborne nitrate levels in Colombia (Cuello et al., 1976) and in the Nottinghamshire, UK, mining town of Worksop (Hill et al., 1973). In Chile, nitrate fertilizer usage was taken as an estimate of population exposure to nitrate, whereas in Colombia and Worksop high urinary nitrate concentrations were assumed to reflect high nitrate intake. Nitrite concentrations were also measured in gastric juice in high-risk areas of Colombia (Tannenbaum et al., 1979). In contrast, no measurement of nitrate or nitrite was made in case-control studies in Japan (Haenszel et al., 1976a) or in Japanese Hawaians (Haenszel et al., 1972), in which gastric cancer risk was related to the consumption of certain foods and to well-water use in Japan.

The principal features of the early studies in South America and Japan are summarized in the review by the National Academy of Sciences (1981) of the health effects of nitrate, nitrite and $N$-nitroso compounds. In the absence of reliable measures of nitrate intake in populations with different risks, the evidence for a role of dietary nitrate in the etiology of gastric cancer is inconclusive. I do not propose to elaborate on these early studies but will concentrate instead on the results of more recent investigations, in the first instance with respect to gastric cancer and nitrate levels in drinking-water. Reports of studies in the UK (Davies, 1980; Fraser & Chilvers, 1981; Beresford, 1981), Chile (Zaldivar & Wetterstrand, 1978), Hungary (Juhász et al., 1980), Italy (Amadori et al., 1980) and Denmark (Jensen, 1982) have appeared in the last few years.

## STOMACH CANCER AND NITRATE IN DRINKING-WATER

### England and Wales

In the UK, much publicity has been given recently to drinking-water as a source of nitrate. The media are quick to comment on the increase in the use of nitrogenous fertilizers in this country over the last 30 years, and the rise in waterborne nitrate levels in the last 20 years. In the haste to report these increasing trends, the fact that England and Wales, in common with many other areas, have experienced a marked decline in gastric cancer mortality is often overlooked. Death rates have fallen over the last 30 years among people of all ages and of both sexes (Office of Population Censuses and Surveys and Cancer Research Campaign, 1981). In fact, with the exception of the oldest age group, the decline in women began as early as 50 years ago, and in men about a decade later.

Age-specific incidence rates, available nationally for only the last 20 years, are also falling, except for people over the age of 75 (Cancer Research Campaign, 1982). In this oldest age group, an increase in 1974 probably reflects improved ascertainment of stomach cancer following the introduction of a simplified national scheme for cancer registration, rather than a genuine increase in incidence. The prognosis for patients with stomach cancer remains poor, only 7% of registered cases surviving five years. The substantial decline in mortality is not due therefore to an improvement in survival.

The use of nitrogenous fertilizers has increased markedly over the last 30 years, with application rates (in 1970–1972) ranging from 28 kg per hectare in South Wales to 68 in East Anglia (Central Water Planning Unit, 1977). It is widely assumed that nitrate pollution of water supplies stems from unused fertilizer nitrogen. However, recent studies using $N^{15}$-labelled fertilizer have shown that most of the nitrate lost by leaching comes from the organic reserves in the soil and not directly from the fertilizer by simple leaching as had been supposed (Owen & Dowdell, personal communications). It is only when nitrogen fertilizer applications greatly exceed the economic optimum application rates, or are applied to cereal crops at the wrong time, that leaching of nitrate from fertilizer may be a problem. The soil nitrogen cycle generates nitrate throughout the spring, summer and autumn, and it is mostly this nitrate that is leached out in the autumn and winter months, or released when a pasture is converted to arable land and ploughed intensively.

Whatever the origin of waterborne nitrate from farming, fertilizer usage gives an indication of the level of agricultural activity, and it is therefore of interest to examine regional trends in stomach cancer mortality in relation to fertilizer usage. Clearly, given the existence of opposing time trends, the relationship, if any, is not straightforward. The decrease in stomach cancer mortality is apparent in all regions of England and Wales, and there is no association between the rate of decline and the rate of increase in fertilizer usage (Fraser & Chilvers, 1981). Furthermore, there is a clear inverse relationship between cumulative fertilizer usage (from 1938–1972) and stomach cancer mortality (in 1969–1973) in the rural aggregates of the standard regions, the traditionally agricultural areas in the south and east of England, with higher cumulative fertilizer usage experiencing lower mortality.

Bearing in mind that the movement of nitrate from the land surface to underground water sources is slow, and that there is also a time interval between exposure and death, some may argue that it is still too soon for the full impact of the intensive agricultural activity of recent years to be manifest. The trends in health statistics certainly give no cause for concern at present and provide no suggestion that the increasing use of nitrogenous fertilizers plays any role in the etiology of stomach cancer in England and Wales.

One of the earliest studies of the relationship between nitrate in drinking-water and stomach cancer was carried out by Hill *et al.* (1973) in Worksop, a Nottinghamshire mining town where for many years the public water supply had contained an average of 90 mg/l nitrate—well above the level of 50 mg/l recommended by the WHO (1970), but still a level regarded as acceptable for drinking purposes. Hill *et al.* suggested that, by comparison with national rates, mortality from stomach cancer in Worksop in 1963–1971 was abnormally high in women (observed : expected ratio, 1.60) and higher than that in nine neighbouring towns supplied with low-nitrate water. One of these (Chesterfield) also had a mortality ratio significantly above the national average.

Male mortality in Worksop (observed : expected, 1.08) was similar to that in several other towns and was not significantly increased.

These preliminary observations were much quoted, until Davies (1980) made a more detailed study of Worksop and other mining and non-mining towns in Nottinghamshire. She had access to revised population estimates and to data over a longer period of time, and she adjusted for differences in social class distribution and the proportion of miners when calculating standardized mortality ratios (SMR). The SMR for females in Worksop decreased in significance with each adjustment until, finally, although still raised at 131, it did not differ significantly from the national average. Similar results were obtained when examining deaths occurring during 1958–1975. Thus, Davies concluded that, if allowance is made for social class structure and the number of miners in each town, there is little indication that people of either sex in Worksop have a higher death rate from stomach cancer than those in neighbouring mining towns.

Our own correlation studies of stomach cancer mortality in 1969–1978 in 32 rural districts in eastern England in relation to nitrate concentrations in the public water supplies since about 1955 yielded inconsistent results (Fraser & Chilvers, 1981). In the Anglian Water Authority area we demonstrated a significant trend in male gastric cancer mortality with increasing concentrations of waterborne nitrate, but the trend diminished in strength over time and was not apparent in females. While domestic water supplies seem an unlikely explanation for these findings, the trend in males could not be explained in terms of differences in social class distribution, nor in the proportion of miners, chemical workers or agricultural workers in each nitrate category. Male mortality in rural districts in the Yorkshire Water Authority area followed a similar pattern, but the difference in gastric cancer mortality between the nitrate categories was not significant, and the findings in females were inconsistent over time.

While intensive agricultural activity is the major factor responsible for rising nitrate levels in underground water sources, the increased recycling of sewage effluent is a contributory factor in lowland rivers. Beresford (1981) studied the relationship between re-use of water and hazards to health in the London area; nitrate was one of several indicators used to assess the degree of re-use. Mortality from different causes, principally cancer, was examined for 29 London boroughs for the period 1968–1974. Socio-economic characteristics of the boroughs were found to account for the statistical associations between water re-use and each cause studied, except male stomach cancer, for which a weak residual association remained. Even this association disappeared when variation in the size of the boroughs was taken into account. In a more extensive study of 253 UK towns, Beresford (personal communication) found no evidence of a positive association between nitrate levels in drinking-water and mortality from all cancers, or from stomach cancer in particular.

### Chile

Chile has natural deposits of nitrate, and for decades large amounts of these compounds have been used as fertilizers. Stomach cancer is very common in people of both sexes, and there is a marked geographical variation in mortality, which several workers have correlated with nitrate fertilizer usage (Zaldivar & Robinson, 1973; Armijo & Coulson, 1975; Zaldivar & Wetterstrand, 1975; Zaldivar, 1977). Zaldivar and Wetterstrand (1978) also examined nitrate levels in drinking-water supplying 202 urban areas in 25 provinces in relation to corresponding age-adjusted stomach cancer death rates. The nitrate levels ranged from 0–30 ppm nitrate nitrogen (0–133 mg/l nitrate) with a mean of 1.446 (6.4 mg/l nitrate). Only two provinces reported nitrate nitrogen levels above 11.3 ppm, the WHO recommended level on this scale.

Zaldivar and Wetterstrand found no association between the nitrate levels in provincial water supplies and death rates from stomach cancer in people of either sex. When the 25 provinces were aggregated into six geographical areas, similar non-significant correlations were found ($r = 0.1367$ and $0.1143$ in males and females, respectively). The authors suggested that the rural population, who drink water mainly from natural springs and artesian wells, may be exposed to higher waterborne nitrate levels than urban dwellers, but no data were available for analysis. The results they presented provide no evidence that the high stomach cancer mortality in Chile is related to nitrate levels in drinking-water.

## Hungary

Hungary is another country where there is high mortality from stomach cancer and where shallow wells often contain high nitrate levels. An epidemiological study was started in 1975 to examine the relationship between drinking-water nitrate, methaemoglobinaemia, soil type and stomach cancer in 230 localities in the county of Szabolcs-Szatmar (Juhász et al., 1980). Each locality was allocated to one of four groups according to stomach cancer incidence ($\leq 20$ per 100 000 or $> 20$) and well-water nitrate concentration ($\leq 100$ ppm nitrate or $> 100$).

A high incidence of stomach cancer was found in 60% of all localities in the county, and the majority of these (127 of 139) had high concentrations of nitrate in the drinking-water; 78 out of 91 localities with a low incidence also had high-nitrate water. The finding of 13 localities where both gastric cancer incidence and nitrate levels were low, and of the 127 localities where both were high, is consistent with a role for nitrate in the etiology of stomach cancer. The 12 localities where nitrate levels were low but gastric cancer incidence was high clearly are not, and nor are the 78 localities with a low incidence yet high nitrate levels in drinking-water. However, the mean urinary nitrate concentration was not high in this latter category in a random sample of single urine specimens, suggesting that, despite high levels in drinking-water, total nitrate intake may not have been high. The small size of many of the localities and absence of information on the variability of the well-water nitrate levels cast doubt on the reliability of both the incidence rates and the waterborne nitrate levels in this study.

## Italy

The publication by Amadori et al. (1980) of some preliminary data on a group of 92 Italian farmworkers received wide publicity when it was claimed that well-water nitrate levels of 44 ppm were causing stomach cancer in these workers. This cannot be inferred from the data presented in the paper, for no information is given on the levels of nitrate in the drinking-water of the urban population at lower risk with whom the farmworkers were being compared. Furthermore, the farmworkers were heavily exposed to agricultural chemicals, including carbamates, which can react with nitrites in the soil and crops to form $N$-nitroso compounds. Mean salivary nitrate and nitrite levels were only slightly higher than levels found in the UK (Forman, personal communication), but 42 of the 46 farmworkers who underwent gastroscopy had superficial or chronic atrophic gastritis. The authors recognized the need for similar studies in individuals in low-risk areas for stomach cancer.

## Denmark

Much publicity too has been given to Jensen's (1982) finding of a higher incidence of stomach cancer in Aalborg, with an average of 30 mg/l nitrate in its drinking-water, compared with Aarhus, with low-nitrate water supplies. Stomach cancer decreased markedly in both towns over the 30-year period covered by the study, while nitrogen fertilizer usage in Denmark has increased, and the consumption of vitamin C, the best-known nitrosation inhibitor, has remained virtually constant. The cancer pattern seen in Aalborg is compatible with an assumption of a socioeconomic status lower than that in Aarhus, but a comparison of available social class indicators revealed no obvious difference between the towns.

Urinary nitrate concentrations in early-morning specimens were measured in 42 and 43 children aged 11–14 years in two school classes in Aalborg and Aarhus, respectively. The distribution in Aalborg was skewed towards higher values, but there was no significant difference between the means (0.42 mmol/l and 0.28 mmol/l) or median values (0.33 mmol/l and 0.26 mmol/l) in the two towns. While Jensen suggested tentatively that his results support a possible weak role for nitrate in the etiology of stomach cancer, lack of definite evidence of higher nitrate intake in Aalborg weakens this conclusion. In fact, lack of information on total nitrate intake in populations at differing risk, or individuals with and without stomach cancer, is a weakness of many epidemiological studies in this area.

## STOMACH CANCER AND TOTAL NITRATE INTAKE

While concern is often expressed over rising levels of nitrate in drinking-water, consideration is seldom given to the contribution made by water to total nitrate intake. Vegetables and cured meat products are important sources of nitrate in the adult diet, and milk and cheese, fats, fish, fresh meat and offal, sugar and preserves also contain small quantities of nitrate.

We have recently completed a survey of dietary nitrate in well-water users, which has provided some information on the relative importance of water and food as sources of nitrate (Chilvers et al., 1984). The well-water nitrate levels ranged from 0 to 269 mg/l, and many wells had substantially higher levels than the public water supplies in the same area. Because nitrate levels in single urine specimens and saliva vary widely during the day, 24-hour urinary nitrate excretion was used as a more reliable surrogate measure of total nitrate intake (Ellen et al., 1982). From this estimate and an assay of the nitrate content of drinking-water, in conjunction with a diary record of consumption, it was possible to obtain an estimate of nitrate intake from water. The diet diary was also used to provide a rough estimate of nitrate intake from solid food. The contributions of waterborne nitrate and food to total nitrate intake were assessed from these data over a wide range of concentrations of nitrate in drinking-water.

When the waterborne nitrate level was less than 50 mg/l, as recommended by the WHO, 30% of ingested nitrate was from water. As the well-water nitrate concentration rose, the contribution of water to daily intake increased; at levels between 50 and 100 mg/l, on average, nearly 70% of daily intake was from water, and above 100 mg/l over 80% of daily intake was waterborne. Thus, it is only at levels well above those currently recommended that waterborne nitrate is the major contributor to total nitrate intake.

There was a consistent rise in 24-hour urinary nitrate excretion with increasing concentrations of well-water nitrate, whatever the intake of nitrate from food. The mean level over all food categories varied from 86 mg when nitrate-free water was consumed to 260 mg when the well-water nitrate levels exceeded 100 mg/l. The mean level was 118 mg when the well-water nitrate concentration was in the range usually found in public water supplies in the UK.

Public speculation over the nitrate content of vegetables in relation to increasing inorganic fertilizer usage prompted an examination of the difference in urinary nitrate excretion between well-water users who ate only commercially-produced vegetables and those who grew some of their own produce. There was no difference in urinary excretion, and therefore in intake, between the two groups.

There was no consistent relationship between nitrate intake and social class in this survey, but studies in progress in Oxford have demonstrated an association between social class and salivary nitrate concentrations with higher levels in the higher social classes (Forman, personal communication). Since gastric cancer exhibits a steep social class gradient, with higher mortality in the lower social classes (Office of Population Censuses and Surveys and Cancer Research Campaign, 1981), this trend in salivary nitrate is the opposite of that expected if nitrate intake were an important determinant of gastric cancer risk.

*Chile*

Following the demonstration of a strong statistical association between nitrate fertilizer usage and gastric cancer mortality in Chile (Zaldivar & Robinson, 1973; Armijo & Coulson, 1975; Zaldivar & Wetterstrand, 1975; Zaldivar, 1977), a case-control study was carried out to test the association and to look for other etiological factors (Armijo et al., 1981a). The study showed that stomach cancer was associated with previous occupation in agriculture, and that patients had resided in high-risk areas during early life for longer periods than had controls. However, more detailed studies of nitrate intake revealed significantly higher nitrate levels in the urine of school children and in vegetables in a low-risk area (Armijo et al., 1981b). Salivary nitrite levels were similar in all four areas studied, but nitrite levels in vegetables were inconsistent. Explanations are now being sought for these paradoxical findings, which do not support the hypothesis that high nitrate ingestion is involved in the etiology of stomach cancer.

## China

By contrast with these latest results from Chile, preliminary data from China show that levels of nitrate and nitrite in vegetables and drinking-water are higher in areas of high risk of stomach cancer than in low-risk areas (Xu, 1981). Nitrate and nitrite levels in saliva and gastric juice of fasting patients with chronic gastritis were also found to be higher in high-risk than in low-risk areas, with the incidence of chronic gastritis running parallel to gastric cancer mortality rates.

## Colombia

In Colombia, too, where the evidence for a link between stomach cancer and high nitrate ingestion is most persuasive, there is a high prevalence of gastric lesions, such as superficial gastritis, chronic atrophic gastritis and intestinal metaplasia (Correa et al., 1970, 1976; Haenszel et al., 1976b). These recognized precursors of stomach cancer are associated with low gastric acidity and high gastric nitrite levels (Tannenbaum et al., 1979), and they are most prevalent in the impoverished communities in Colombia where stomach cancer is very common and the major cause of death. In gastroscopic studies among volunteers in a high-risk area, 75% were found to have some form of gastritis by 25 years of age (Correa et al., 1976). A high corn diet was found to be associated with gastric lesions, but ingestion of lettuce, which contains vitamin C, was inversely related (Haenszel et al., 1976b).

Wells with nitrate concentrations up to 300 mg/l are a feature of several high-risk area in Colombia, and the high urinary nitrate levels found in single specimens (Cuello et al., 1976) and 12-hour collections (Shaheen, personal communication) suggest that nitrate intake is also high. The high urinary levels were not confined to drinkers of nitrate-rich well-water, for high levels were found in an area with nitrate-free water supplies, suggesting that locally-grown food, rather than water was the nitrate source.

Following these studies in Colombia, several investigators have now demonstrated an association between high gastric nitrite levels, low acidity and the presence of $N$-nitroso compounds (Reed et al., 1981). In 301 gastric juice samples from 267 untreated subjects, including 50 healthy volunteers, $N$-nitrosamine concentrations rose progressively with pH, patients with conditions associated with hypochlorhydria showing correspondingly higher mean levels. Thus, patients with chronic gastritis, gastric ulcer, a partial gastrectomy, pernicious anaemia and stomach cancer itself, in whom the pH was greater than 3, had higher $N$-nitrosamine levels than normal subjects and patients with conditions associated with more acid stomach contents.

An increased risk of stomach cancer in conditions associated with low gastric acidity is well-recognized, and results of this nature lend support to the hypothesis that $N$-nitroso compounds may be involved in its development. Their formation can be readily inhibited in vitro and in vivo by antioxidants such as vitamin C, and Reed et al. (1983) have demonstrated for the first time in humans a significant lowering of

*N*-nitroso compounds in gastric juice by ascorbate treatment in 51 achlorhydric subjects. Clearly, this observation may have important implications for the prevention of stomach cancer in high-risk subjects should *N*-nitroso compounds be shown to be causative agents.

## OESOPHAGEAL CANCER

*Iran*

The fact that many *N*-nitroso compounds have produced oesophageal tumours in laboratory animals, led the IARC to embark on a search for *N*-nitrosamines and their precursors as possible causative agents of oesophageal cancer in north-east Iran (Joint Iran/IARC Study Group, 1977). The striking finding in these detailed investigations was the severely limited and probably irritant nature of the diet in this high-risk area. The level of nitrate and nitrite in water, wheat and bread was not particularly high, and there was no significant difference in the average daily intake of nitrate or nitrite between high- and low-incidence areas. Typical dietary items were analysed for volatile nitrosamines, polycyclic aromatic hydrocarbons and aflatoxins, but the results showed no unusual level of any of these potential carcinogens.

*China*

In contrast, Chinese scientists have suggested that *N*-nitrosamines are likely to be causative agents in their country (Yang, 1980). A diet rich in nitrates, nitrites, secondary amines and *N*-nitrosamines, but low in vitamin C is a feature of high-risk areas, where oesophageal dysplasia is also prevalent. A great deal of information on oesophageal cancer and possible risk factors has been gathered in China, but, while *N*-nitrosamines are among the prime suspects, it has not been established that they are the causative agents.

## CONCLUSION

In my view, the evidence from epidemiological studies that dietary nitrate is involved in the etiology of stomach cancer in developed countries is very weak. In the UK, and elsewhere, in spite of an increase in exposure through rising nitrate levels in drinking-water, and possibly in vegetables, stomach cancer is declining steadily.

None of the recent studies that examined the relationship between stomach cancer and nitrate in drinking-water provides convincing evidence of a link. The studies by Davies and Beresford in the UK, and by Zaldivar and Wetterstrand in Chile gave negative results; our findings in the eastern UK and those of Juhász *et al.* in Hungary are inconsistent; and no inference can be drawn from the study of Italian farmworkers in the absence of a low-risk group for comparison. Jensen's study in Denmark lacks definite evidence of a difference in nitrate intake between the two towns. With the exception of the Hungarian study, nitrate levels in these studies rarely exceeded the WHO recommended level of 50 mg/l, so that the contribution of water to daily nitrate intake would be no more than one-third in persons consuming a normal western diet, and food would be the more important source of nitrate in the populations studied.

The conflicting results from Chile are difficult to interpret, but the evidence from impoverished communities in Colombia and China, where conditions associated with low gastric acidity and dietary deficiencies are common, suggests that high nitrate ingestion may be one of several factors that may interact to increase the risk of stomach cancer in these populations. Similarly, the evidence suggests that an interaction between nitrate and many other factors may contribute to the high incidence of oesophageal cancer in parts of China, though not in Iran.

The demonstration of significantly raised concentrations of $N$-nitroso compounds in the gastric juice of patients with conditions associated with or conducive to gastric cancer lends support to the view that these compounds may be involved in gastric carcinogenesis. However, their presence does not establish their role as causal.

## ACKNOWLEDGEMENTS

The author is supported by the Medical Research Council.

## REFERENCES

Amadori, D., Ravaioli, A., Gardini, A., Liverani, M., Zoli, W., Tonelli, B., Ridolfi, R. & Gentilini, P. (1980) N-Nitroso compound precursors and gastric cancer: preliminary data of a study on a group of farm workers. *Tumori,* **66,** 145–152

Archer, M.C., Lee, L. & Bruce, W.R. (1982) *Analysis and formation of nitrosamines in the human intestine.* In: Bartsch, H., O'Neill, I.K., Castegnaro, M. & Okada, M., eds, N-*Nitroso Compounds: Occurrence and Biological Effects (IARC Scientific Publications No. 41),* Lyon, International Agency for Research on Cancer, pp. 357–363

Armijo, R. & Coulson, A.H. (1975) Epidemiology of stomach cancer in Chile—the role of nitrogen fertilizers. *Int. J. Epidemiol.,* **4,** 301–309

Armijo, R., Orellana, M., Medina, E., Coulson, A.H., Sayre, J.W. & Detels, R. (1981a) Epidemiology of gastric cancer in Chile: I. Case-control study. *Int. J. Epidemiol.,* **10,** 53–56

Armijo, R., Gonzalez, A., Orellana, M., Coulson, A.H., Sayre, J.W. & Detels, R. (1981b) Epidemiology of gastric cancer in Chile: II. Nitrate exposures and stomach cancer frequency. *Int. J. Epidemiol.,* **10,** 57–62

Beresford, S.A.A. (1981) The relationship between water quality and health in the London area. *Int. J. Epidemiol.,* **10,** 103–115

Brooks, J.B., Cherry, W.B., Thalker, L. & Alley, C.C. (1972) Analysis by gas chromatography of amines and nitrosamines produced *in vivo* and *in vitro* by *Proteus mirabilis. J. infect. Dis.,* **126,** 143–153

Cancer Research Campaign (1982) *Trends in Cancer Survival in Great Britain,* London

Central Water Planning Unit (1977) *Nitrate and Water Resources with Particular Reference to Groundwater,* Reading

Chilvers, C., Inskip, H., Caygill, C., Bartholomew, B., Fraser, P. & Hill, M. (1984) A survey of dietary nitrate in well-water users. *Int. J. Epidemiol., 13,* 324-331

Correa, P., Cuello, C. & Duque, E. (1970) Carcinoma and intestinal metaplasia of the stomach in Colombian migrants. *J. natl Cancer Inst., 44,* 297-306

Correa, P., Cuello, C., Duque, E., Burbano, L.C., Garcia, F.T., Bolanos, O., Brown, C. & Haenszel, W. (1976) Gastric cancer in Colombia. III. Natural history of precursor lesions. *J. natl Cancer Inst., 57,* 1027-1033

Cuello, C., Correa, P., Haenszel, W., Gordillo, G., Brown, C., Archer, M. & Tannenbaum, S. (1976) Gastric cancer in Colombia. I. Cancer risk and suspect environmental agents. *J. natl Cancer Inst., 57,* 1015-1020

Davies, J.M. (1980) Stomach cancer mortality in Worksop and other Nottinghamshire mining towns. *Br. J. Cancer, 41,* 438-445

Ellen, G., Schuller, P.L., Bruijns, E., Froeling, P.G.A.M. & Baadenhuijsen, H. (1982) *Volatile N-nitrosamines, nitrate and nitrite in urine and saliva of healthy volunteers after administration of large amounts of nitrate.* In: Bartsch, H., O'Neill, I.K., Castegnaro, M. & Okada, M., eds, N-*Nitroso Compounds: Occurrence and Biological Effects (IARC Scientific Publications No. 41),* Lyon, International Agency for Research on Cancer, pp. 365-378

Fraser, P. & Chilvers, C. (1981) Health aspects of nitrate in drinking water. *Sci. total Environ., 18,* 103-116

Fraser, P., Chilvers, C., Beral, V. & Hill, M.J. (1980) Nitrate and human cancer: a review of the evidence. *Int. J. Epidemiol., 9,* 3-11

Haenszel, W., Kurihara, M., Segi, M. & Lee, R.K.C. (1972) Stomach cancer among Japanese in Hawaii. *J. natl Cancer Inst., 49,* 969-988

Haenszel, W., Kurihara, M., Locke, F.B., Shinuzu, K. & Segi, M. (1976a) Stomach cancer in Japan. *J. natl Cancer Inst., 56,* 265-274

Haenszel, W., Correa, P., Cuello, C., Guzman, N., Burbano, L.C., Lores, H. & Munoz, J. (1976b) Gastric cancer in Colombia. II. Case-control epidemiologic study of precursor lesions. *J. natl Cancer Inst., 57,* 1021-1026

Hicks, R.M., Walters, C.L., Elsebai, I., El-Aassar, A.-B., El Merzebani, M. & Gough, T.A. (1977) Demonstration of nitrosamines in human urine: preliminary observations on a possible etiology for bladder cancer in association with chronic urinary tract infection. *Proc. R. Soc. Med; 70,* 413-417

Hill, M.J., Hawksworth, G. & Tattersall, G. (1973) Bacteria, nitrosamines and cancer of the stomach. *Br. J. Cancer, 28,* 562-567

Jensen, O.M. (1982) Nitrate in drinking water and cancer in Northern Jutland, Denmark, with special reference to stomach cancer. *Ecotoxicol. environ. Saf., 6,* 258-267

Joint Iran/IARC Study Group (1977) Oesophageal cancer studies in the Caspian littoral of Iran: results of population studies—a prodrome. *J. natl Cancer Inst., 59,* 1127-1138

Juhász, L., Hill, M.J. & Nagy, G. (1980) *Possible relationship between nitrate in drinking water and incidence of stomach cancer.* In: Walker, E.A., Castegnaro, M., Griciute, L. & Börzsönyi, M., eds, N-*Nitroso Compounds: Analysis, Formation and Occurrence (IARC Scientific Publications No. 31),* Lyon, International Agency for Research on Cancer, pp. 619-623

National Academy of Sciences (1981) *The Health Effects of Nitrate, Nitrite and N-Nitroso Compounds,* Washington DC, National Academy Press

Office of Population Censuses and Surveys and Cancer Research Campaign (1981) *Cancer Statistics: Incidence, Survival and Mortality in England and Wales (Studies on Medical and Population Subjects No. 43),* London, Her Majesty's Stationery Office

Radomski, J.L., Greenwald, D., Hearn, W.L., Block, N.L. & Woods, F.M. (1978) Nitrosamine formation in bladder infection and its role in the etiology of bladder cancer. *J. Urol.,* **120,** 48–50

Reed, P.I., Smith, P.L.R., Haines, K., House, F.R. & Walters, C.L. (1981) Gastric juice $N$-nitrosamines in health and gastroduodenal disease. *Lancet, ii,* 550–552

Reed, P.I., Summers, K., Smith, P.L.R., Walters, C.L., Bartholomew, B.A., Hill, M.J., Vennitt, S., Hornig, D. & Bonjour, J.P. (1983) Effect of ascorbic acid treatment on gastric juice nitrite and $N$-nitroso compound concentrations in achlorhydric subjects. *Gut,* **24,** A 492–493

Royal Commission on Environmental Pollution (1979) *7th Report, Agriculture and Pollution,* Chapter IV, London, Her Majesty's Stationery Office, pp. 87–125

Tannenbaum, S.R. (1983) N-Nitroso compounds: a perspective on human exposure. *Lancet, i,* 629–632

Tannenbaum, S.R., Archer, M.C., Wishnok, J.S. & Bishop, W.W. (1978) Nitrosamine formation in human saliva. *J. natl Cancer Inst.,* **60,** 251–253

Tannenbaum, S.R., Moran, D., Rand, W., Cuello, C. & Correa, P. (1979) Gastric cancer in Colombia. IV. Nitrite and other ions in gastric contents of residents from a high risk region. *J. natl Cancer Inst.,* **62,** 9–12

Wang, T., Kakizoe, T., Dion, P., Furrer, R., Varghese, A.J. & Bruce, W.R. (1978) Volatile nitrosamines in normal human faeces. *Nature,* **276,** 280–281

WHO (1970) *European Standards for Drinking Water,* 2nd ed., Geneva

Xu, G.-W. (1981) Gastric cancer in China: a review. *J. R. Soc. Med.,* **74,** 210–211

Yang, C.S. (1980) Research on oesophageal cancer in China: a review. *Cancer Res.,* **40,** 2633–2644

Zaldivar, R. (1977) Nitrate fertilizers as environmental pollutants: positive correlation between nitrates ($NaNO_3$ and $KNO_3$) used per unit area and stomach cancer mortality rates. *Experientia,* **33,** 264–265

Zaldivar, R. & Robinson, H. (1973) Epidemiological investigation on stomach cancer mortality in Chileans: association with nitrate fertilizer. *Z. Krebsforsch.,* **80,** 289–295

Zaldivar, R. & Wetterstrand, W.H. (1975) Further evidence of a positive correlation between exposure to nitrate fertilizers ($NaNO_3$ and $KNO_3$) and gastric cancer death rates: nitrites and nitrosamines. *Experientia,* **31,** 1354–1355

Zaldivar, R. & Wetterstrand, W.H. (1978) Nitrate nitrogen levels in drinking water of urban areas with high- and low-risk populations for stomach cancer: an environmental epidemiology study. *Z. Krebsforsch.,* **92,** 227–234

# NITRATES:

# CONCLUSION

Rapporteur: B.K. ARMSTRONG

Nitrates are unique among the agents discussed in this volume in that there is no direct laboratory evidence to support the notion that they are carcinogenic. The concern is that, through in-vivo conversion to nitrite, they may result in increased endogenous formation of $N$-nitroso compounds. Many $N$-nitroso compounds are carcinogenic in experimental animals, although there is no direct evidence of their carcinogenicity to man. Patterns of cancer incidence typical of the relevant $N$-nitroso compounds can be induced in experimental animals fed large amounts of nitrite and the appropriate secondary amines. Thus, at least under these circumstances, in-vivo nitrosation reactions can lead to carcinogenesis.

Studies of human cancer in relation to nitrate exposure were stimulated by the possibility that it might contribute to human carcinogenesis by the mechanism described above. The studies to date have been almost entirely descriptive, relating incidence or mortality from cancer in whole populations to various measures of nitrate intake. The results have shown no consistent pattern.

In discussion of the data, the question was raised as to whether the selection of areas for study, such as Chile, Colombia and parts of Europe, had been influenced by a knowledge that high gastric cancer rates and high nitrate intakes coexisted in them. If so, this could explain some of the apparently positive associations. There would be merit in studying the association in new areas where stomach cancer rates are known to be high, cancer incidence or mortality statistics are good and nitrate intake is not currently documented but could be studied.

The difficulties in studying the postulated sequence of dietary nitrate, leading to nitrite, $N$-nitroso compounds and cancer in man, were recognized. Descriptive studies are comparatively weak and often uncertain tools for demonstrating causal relationships or for providing firm evidence against them. Analytical studies on this subject have been and will continue to be hampered by difficulty in measuring individual exposure to nitrates over a period potentially relevant to carcinogenesis. A partial answer may lie in intervention studies in groups at high risk of a possibly relevant cancer (e.g., gastric cancer in Colombia) using compounds, such as vitamins C and E and some tannins, known to inhibit in-vivo nitrosation. A positive result in a well-designed study could establish in-vivo nitrosation as a mechanism of carcinogenesis in man; it would not establish necessarily that variation in intake of nitrates, from whatever source, was an important determinant of variation in the degree of in-vivo nitrosation. Whether or not this is true, however, could be established by adequate study of in-vivo nitrosation itself.

There is no evidence that nitrates themselves are carcinogenic, nor any firm laboratory or epidemiological evidence that nitrites, which can be derived from them *in vivo,* cause cancer. However, in view of laboratory evidence that, by acting as nitrosating agents in the formation of *N*-nitroso compounds, nitrites can be contributory factors in experimental carcinogenesis, further human studies are warranted. The epidemiological evidence currently available is qualitatively inadequate to justify any useful conclusions on the role of *N*-nitroso compounds or nitrates in human carcinogenesis.

# BERYLLIUM

BERRYLLIUM:

# LABORATORY EVIDENCE

W.G. FLAMM

*Office of Toxicological Sciences, Bureau of Foods, Food and Drug Administration, Washington DC, USA*

## INTRODUCTION

Beryllium and a series of beryllium compounds have been subjected to a large number of carcinogenicity studies in animals. The compounds include passivated beryllium metal, beryllium hydroxide and oxide, salts of beryllium (sulfate, silicate, phosphate) and beryllium alloys (BeAl, BeCu, BeCuCo, BeNi, $Be_xB$, $VBe_{12}$, $TiBe_{12}$, $TaBe_{12}$, $NbBe_{12}$). The carcinogenicity of beryllium compounds in laboratory animals has been known since 1946 (Gardner & Heslington, 1946) and has been reviewed by Schepers (1961), Sunderman (1971), Infante and Wagoner (1975), Sunderman (1977), Reeves (1978) and Groth (1980). Beryllium and beryllium-containing compounds have been tested and found to be carcinogenic by oral administration, inhalation, intratracheal administration, intravenous injection and intramedullary injection into bone. Since 1946, approximately 30 studies have been reported on various aspects of the carcinogenicity of 13 different beryllium compounds (IARC, 1980).

## RESULTS

Beryllium compounds were the first nonradioactive chemicals found to cause osteogenic sarcoma in animals. Gardner and Heslington (1946) found that zinc beryllium silicate and beryllium oxide administered intravenously induced osteogenic sarcomas in rabbits. This discovery touched off a new era in the study of inorganic compounds, and particularly beryllium-containing compounds. In addition to silicates and oxides, beryllium metal, and ZnMnBeSiOx have been demonstrated to induce osteogenic sarcoma following intravenous injection.

Vorwald (1953) showed that rats exposed by inhalation to beryllium sulfate for more than a year developed pulmonary adenocarcinomas. These findings were confirmed by Schepers (1955), who found a 39% incidence of pulmonary tumours,

10% of which had metastasized, following inhalation exposure (six months) to beryllium sulfate. Lung cancers were found to metastasize to adrenals, kidney, liver, pancreas and brain (Schepers et al., 1957). Beryllium ore containing 4% beryllium has also been shown to induce pulmonary adenomas, adenocarcinomas and epidermoid carcinomas in rats (Wagner et al., 1969). Inhalation of other beryllium-containing compounds has also been reported to induce lung cancer in rats. These compounds include the following: beryllium fluoride, zinc manganese beryllium silicate and beryllium phosphate, beryllium oxide and hydroxide. Beryllium is by far the most potent inorganic pulmonary carcinogen tested in rats, being a thousand times more active per unit weight than chrysotile in producing lung tumours.

Beryllium salts have not been shown to be active in short-term mutagenicity tests, nor do salts of beryllium appear to form covalent bonds in DNA. Beryllium salts may, however, interact with DNA in other ways, and they have been reported to cause chromosomal aberrations in cultures of certain mammalian cells.

## SUMMARY

Beryllium-containing compounds have been studied extensively and have been known to be carcinogenic in animals since 1946. Beryllium salts and alloys were among the first nonradioactive, inorganic substances shown to induce osteogenic sarcoma in experimental animals. Beryllium-containing compounds have been demonstrated to be powerful pulmonary carcinogens in rats. To date, these compounds do not appear to be mutagenic, leaving open the question of their mechanism of action.

## REFERENCES

Gardner, L.U. & Heslington, H.F. (1946) Osteosarcoma from intravenous beryllium compounds in rabbits. *Fed. Proc., 5,* 221

Groth, D.H. (1980) Carcinogenicity of beryllium: review of the literature. *Environ. Res., 21,* 56–62

IARC (1980) *IARC Monographs on the Evaluation of the Carcinogenic Risk of Chemicals to Humans,* Vol. 23, *Some Metals and Metallic Compounds,* Lyon, pp. 143–204

Infante, P.F. & Wagoner, J.K. (1975) *Evidence for the carcinogenicity of beryllium.* In: Hutchinson, T.C., ed., *Proceedings of the International Conference on Heavy Metals in the Environment,* Toronto, Institute for Environmental Studies, pp. 329–338

Reeves, A.L. (1978) *Beryllium carcinogenesis.* In: Schrauzer, G.N., ed., *Inorganic and Nutritional Aspects of Cancer,* New York, Plenum Press, pp. 13–27

Schepers, G.W.H. (1955) *Recent observations on chronic pulmonary beryllium disease.* In: *Transactions of the 20th Annual Meeting of Industrial Hygiene Foundation,* pp. 139–156

Schepers, G.W.H., Durkan, T.M., Delahant, A.V. & Creedon, F.T. (1957) The biological action of inhaled beryllium sulfate. *A.M.A. Arch. ind. Health, 15,* 32–58

Schepers, G.W.H. (1961) Neoplasia experimentally induced by beryllium compounds. *Prog. exp. Tumor Res., 2,* 203–244

Sunderman, F.W., Jr (1971) Metal carcinogenesis in experimental animals. *Food Cosmet. Toxicol., 9,* 105–120

Sunderman, F.W., Jr (1977) Metal carcinogenesis. *Adv. mod. Toxicol., 1,* 257–295

Vorwald, A.J. (1953) *Adenocarcinoma in the lung of albino rats exposed to compounds of beryllium.* In: Cancer of the Lung, An Evaluation of the Problem *(Proc. Sci. Session, Am. Cancer Soc. Ann. Meet.),* New York, American Cancer Society, pp. 103–109

Wagner, W.D., Groth, D.H., Holtz, J.L., Madden, G.E. & Stokinger, H.E. (1969) Comparative chronic inhalation toxicity of beryllium ores, bertrandite and beryl, with production of pulmonary tumors by beryl. *Toxicol. appl. Pharmacol., 15,* 10–29

BERYLLIUM:

# EPIDEMIOLOGICAL EVIDENCE

R. SARACCI

*Unit of Analytical Epidemiology, Division of Epidemiology and Biostatistics, International Agency for Research on Cancer, Lyon, France*

## INTRODUCTION

This paper reviews the published epidemiological evidence on the relationship between exposure to beryllium (Be) and cancer occurrence in man. The three initial sections summarize some information—helpful in order to place the question in perspective and to interpret the epidemiological findings—on environmental occurrence and human exposures to Be, its metabolism and its toxic effects in man. The fourth section presents in detail the epidemiological studies that contribute information on the Be/cancer issue. This is followed by a discussion of the epidemiological studies and by a short section of conclusions.

## ENVIRONMENTAL OCCURRENCE AND HUMAN EXPOSURES

(Browning, 1969; Hammond & Beliles, 1980; IARC, 1980; Stokinger, 1981)

Be is a light (atomic weight, 9.01), stable, bivalent metal. Some 50 different minerals contain Be, which is present in a variety of rocks, in most soils, and in coal. Be has been found in samples of air, waters, plants (including tobacco), animal tissues and foods. Production of Be started in the 1920s, expanded during the Second World War, and declined during the 1970s, figures of 320 000 kg in 1969 and 97 100 kg in 1977 being reported for world production outside the USA (IARC, 1980). Be enters as a hardening agent in several alloys, notably with copper, which in 1979 were estimated to absorb some 70–80% of Be employed for industrial uses (15–20% being used as metallic Be and 5–10% as oxide).

Occupational exposures to Be and its derivatives can occur in mining, extraction, refining, alloy manufacture as well as in a number of industrial uses. The latter currently include manufacture of ceramics (Be oxide), the electronics industry (where alloys are utilized for transistors, heat sinks, X-ray and cathode tubes), the nuclear industry (where Be is used as a neutron moderator), and the aerospace industry (where Be is used, e.g., for high-performance brakes). Before 1950, a major use was in the manufacture of fluorescent lights and neon signs.

Atmospheric concentrations in the workplace were high in the past, and levels in the order of 100 or 1000 $\mu g/m^3$ were reported in the years before 1950 from an alloy manufacturing plant (IARC, 1980). These levels are several orders of magnitude higher than those that have produced cancer in experimental animals. Since 1949, exposures in the USA have been drastically reduced, a limit of 2 $\mu g/m^3$ (8-hour time-weighted average concentration), with a ceiling of 25 $\mu g/m^3$, having been recommended by the Atomic Energy Commission. This was widely regarded as a reference criterion until recently.

Environmental exposure to Be derives in part from natural occurrence of the metal. Use by the aerospace industry of Be compounds as propellants provides a possible source of exposure; however, coal combustion is likely to be the main source of environmental contamination. Atmospheric background levels of Be are usually below 0.0001 $\mu g/m^3$, but concentrations 100 times higher have been reported in the neighbourhood of a Be production plant (IARC, 1980).

## METABOLISM

(Hueper, 1966; Browning, 1969; Morgan & Seaton, 1975; Hammond & Beliles, 1980; IARC, 1980; Stokinger, 1981)

Absorption, distribution and excretion depend in part on the molecular species and route of exposure. The metal, which is not absorbed well by any route, may enter the body by inhalation, after deposition in the lungs or through the skin (especially if the latter is not intact), while no appreciable absorption occurs through the gastro-intestinal tract. The metal is probably transported in the bloodstream as an orthophosphate colloid, and, irrespective of the route of administration or the nature of the Be compound, an appreciable portion of the administered dose goes to the skeleton, where it may replace calcium. Other organs mainly involved in the distribution are liver, spleen and kidney. Complete removal of Be deposited in the lungs by inhalation is a slow process, and a small residue may remain for months or years. Similarly, it may take many months or years for the last 5% or so of a dose to be completely cleared from bones and pulmonary lymph nodes. Excretion takes place by the renal route. Concentrations of Be in the lung tissue of occupationally exposed subjects have been found to be on average higher than those in non-exposed people, and a value of 0.02 $\mu g/g$ has been used as a discriminating point to diagnose chronic berylliosis (Sprince et al., 1975).

## TOXIC EFFECTS (OTHER THAN CANCER) IN MAN

(Hueper, 1966; Browning, 1969; Morgan & Seaton, 1975; Hammond & Beliles, 1980; IARC, 1980; Stokinger, 1981)

Three types of pathological effect due to exposure to Be have been well established in man.

### Skin lesions

Itchy rashes, erythematous or papular, can occur on exposed surfaces, often some two weeks after exposure has begun; implantation granulomatas, often chronic, may also occur. Conjunctivitis, sometimes with corneal ulceration, has been reported in exposed workers, and some evidence suggests that hypersensitivity may be involved in skin manifestations, the more soluble Be salts (fluoride, chloride, sulphate tetrahydrate) being the most sensitizing agents.

### Acute berylliosis

This acute syndrome, which has often occurred following high exposures (10–100 $\mu g/m^3$ and above), includes nasopharyngitis (ulcerations and nasal septal perforation may occur), tracheitis with bronchitis and bronchiolitis, which may evolve to a 'chemical' pneumonia. In favourable cases, most of the pathological changes resolve in one to four weeks, but in a proportion of cases (10% or so) chronic berylliosis develops as a sequel.

### Chronic berylliosis (chronic beryllium disease)

This pulmonary and systemic condition may develop sequentially to an acute berylliosis, or as an apparently primary disorder, with an incubation period that may stretch to 10 or more years since the beginning of exposure, which might, however, have ceased several years (5–10) before the disease became clinically apparent. The progression is relatively slow, and a survival time of 15 years is not uncommon. Early case reports, in the mid- and late 1940s, were of workers in the manufacture of fluorescent lamps and of wives of Be production plant workers who inhaled dust from their husbands' clothes. The clinical, radiological and functional features of the disease closely overlap with those of sarcoidosis. The pathological lesion is represented by an interstitial non-caseating granulomatosis of the lung, indistinguishable from sarcoidosis (Sprince et al., 1975). The skin, striated muscles, liver, spleen, kidney and heart may also be affected. No 'dose-response' has been established for this condition, which is interpreted as involving a delayed-type hypersensitivity, so that even very low exposures may be sufficient to induce it (Morgan & Seaton, 1975; Stokinger, 1981; Cotes et al., 1983).

# EPIDEMIOLOGICAL STUDIES OF CANCER IN OCCUPATIONALLY EXPOSED GROUPS

These include a series of investigations by Mancuso (Mancuso & El-Attar, 1969; Mancuso, 1970, 1979, 1980), who followed up at successive dates workers employed at two separate Be extraction, production and fabrication facilities, one in Ohio, the other in Pennsylvania, USA. The latter also formed the object of an investigation by Wagoner et al. (1980). In addition, Infante et al. (1980), with the aim of testing the relationship between cancer risk and Be exposure, analysed the data in the US Beryllium Case Registry, which has been collecting cases of Be-related lung diseases from a wide variety of sources, including the two previously mentioned industrial populations.

*Production plants*

The population investigated in Mancuso's 1979 paper included all white males employed sometime between 1 January 1942 and 31 December 1948 at the facilities in Ohio and Pennsylvania. For cohort identification, the employees payroll record, as submitted quarterly by the employer and kept on microfilm at the Social Security Administration (SSA), were used. The author states (Mancuso, 1970) that these records have 'decided advantages in terms of accuracy and completeness in contrast to the problem of reconstruction of data from personnel files many years later'. Follow-up until 31 December 1974 for the Ohio plant and until 31 December 1975 for the Pennsylvania plant was performed internally by the SSA, matching the cohorts of employees with death claim files. At the Ohio plant, 1222 workers were followed (with 334 deaths), and at the Pennsylvania plant 2044 (with 787 deaths). Causes of death were coded to ICD, 7th Revision, by one state coder. A modified life-table method, using quinquennial time intervals, was employed to calculate person-years and expected deaths, using US national rates, specific for cause-sex-race-calendar period as referent. However, rates for 1965–1967 were also used for 1968–1975. The results with regard to lung cancer (ICD 162, 163) are shown in Table 1 (Ohio plant), Table 2 (Pennsylvania plant) and Table 3 (both plants combined), which present the observed (O) and expected (E) values, their ratios and the 95% confidence intervals (CI) of the ratios. There is a consistent pattern of elevated ($>1$) ratios 15 or more years since first employment: this increase is mostly confined to workers with less than one and with one to four years of employment.

[Note: I have computed ratios and 95% confidence limits using exact confidence limits for the observed values (Documenta Geigy, 1972) and taking the expected values as constants. Table 3 is derived by sum from Tables 1 and 2. To correct for the use of 1965–1967 rates for the period 1968–1975, the expected values as given in the author's paper have been increased by 10% (MacMahon, 1979; Smith, 1981). The same correction was applied to obtain the expected figures in Tables 6 and 9].

In a further expansion of the follow-up (Mancuso, 1980), until 31 December 1976, a total of 3685 workers employed between 1937 and 1948 at the two Be factories were compared to 5929 rayon viscose workers, identified through one company's complete microfilm record of all employment data for individuals employed sometime during

Table 1. Observed (O) and expected (E) deaths due to bronchogenic cancer according to duration of employment and time since onset of employment among white males employed in an Ohio beryllium production facility at some time between 1 January 1942 and 31 December 1948, followed through 31 December 1974

| Interval since onset of employment (years) | Duration of employment (years) | | | | | | | | | Total | | |
|---|---|---|---|---|---|---|---|---|---|---|---|---|
| | <1 | | | 1–4 | | | >5 | | | | | |
| | O/E | Ratio | CI | O/E | Ratio | CI | O/E | Ratio | CI | O/E | Ratio | CI |
| <15 | 3/1.96 | 1.5 | 0.3–4.5 | 0/0.70 | 0 | 0 –5.3 | 0/0.26 | 0 | 0 –14.2 | 3/ 2.92 | 1.0 | 0.2–3.0 |
| ≥15 | 14/7.14 | 2.0 | 1.1–3.3 | 5/1.91 | 2.6 | 0.8–6.1 | 3/1.79 | 1.7 | 0.3– 4.9 | 22/10.84 | 2.0 | 1.3–3.1 |
| Total | 17/9.10 | 1.9 | 1.1–3.0 | 5/2.61 | 1.9 | 0.6–4.4 | 3/2.05 | 1.5 | 0.3– 4.3 | 25/13.76 | 1.8 | 1.2–2.7 |

CI, 95% confidence interval

Table 2. Observed (O) and expected (E) deaths due to bronchogenic cancer according to duration of employment and time since onset of employment among white males employed in a beryllium production facility in Pennsylvania at some time between 1 January 1942 and 31 December 1948, followed through 31 December 1975

| Interval since onset of employment (years) | Duration of employment (years) | | | | | | | | | | | |
|---|---|---|---|---|---|---|---|---|---|---|---|---|
| | <1 | | | 1–4 | | | >5 | | | Total | | |
| | O/E | Ratio | CI | O/E | Ratio | CI | O/E | Ratio | CI | O/E | Ratio | CI |
| <15 | 3/ 4.70 | 0.6 | 0.1–1.9 | 1/2.11 | 0.5 | 0.1–2.6 | 0/0.98 | 0 | 0 –4.1 | 4/ 7.79 | 0.5 | 0.1–1.3 |
| ≧15 | 23/14.12 | 1.6 | 1.0–2.4 | 10/5.80 | 1.7 | 0.8–3.2 | 3/4.30 | 0.7 | 0.1–2.0 | 36/24.22 | 1.5 | 1.0–2.1 |
| Total | 26/18.82 | 1.4 | 0.9–2.0 | 11/7.91 | 1.4 | 0.7/2.5 | 3/5.28 | 0.6 | 0.1–1.7 | 40/32.01 | 1.2 | 0.9–1.7 |

CI, 95% confidence interval

Table 3. Observed (O) and expected (E) deaths due to bronchogenic cancer, Ohio and Pennsylvania facilities pooled

| Interval since onset of employment (years) | Duration of employment (years) | | | | | | | | | | | |
|---|---|---|---|---|---|---|---|---|---|---|---|---|
| | <1 | | | 1–4 | | | >5 | | | Total | | |
| | O/E | Ratio | CI | O/E | Ratio | CI | O/E | Ratio | CI | O/E | Ratio | CI |
| <15 | 6/ 6.66 | 0.9 | 0.3–2.0 | 1/ 2.81 | 0.4 | 0 –1.3 | 0/1.24 | 0 | 0 –3.0 | 7/10.71 | 0.6 | 0.3–1.4 |
| ≥15 | 37/21.26 | 1.7 | 1.2–2.4 | 15/ 7.71 | 2.0 | 1.1–3.2 | 6/6.09 | 1.0 | 0.4–2.4 | 58/35.06 | 1.6 | 1.3–2.1 |
| Total | 43/27.92 | 1.5 | 1.1–2.1 | 16/10.52 | 1.5 | 0.9–2.5 | 6/7.33 | 0.8 | 0.3–1.8 | 65/45.77 | 1.4 | 1.1–1.8 |

CI, 95% confidence interval

Table 4. Lung cancer mortality among beryllium-exposed workers aged 35–74 years as contrasted with that expected on the basis of two cohorts of workers in the viscose rayon industry employed for similar durations of time and followed over the same period of time

| Duration of employment (months) | Lung cancer mortality | | | | | |
|---|---|---|---|---|---|---|
| | O/E[a] | Ratio | CI | O/E[b] | Ratio | CI |
| ≤ 12 | 52/37.60 | 1.4 | 0.9–2.1 | 52/31.67 | 1.6 | 1.1–2.6 |
| 13–48 | 14/13.26 | 1.1 | 0.5–2.2 | 14/10.82 | 1.3 | 0.6–2.9 |
| ≥ 49 | 14/ 6.32 | 2.2 | 0.9–5.7 | 14/ 8.14 | 1.7 | 0.7–4.1 |
| Total | 80/57.18 | 1.4 | 1.0–2.0 | 80/50.63 | 1.6 | 1.1–2.2 |

[a] All viscose rayon employees
[b] Viscose rayon employees who had never transferred from department of initial employment
O, observed; E, expected; CI, 95% confidence interval

1938–1948. Follow-up was implemented by the SSA. For the Be cohort, 1356 deaths occurred (and 20 death certificates could not be located), while among the viscose rayon employees 1824 deaths occurred (78 death certificates could not be located). The results for lung cancer are shown in Table 4, where a comparison is made of the Be workers with all rayon workers and, separately, with rayon workers who never moved out of the department of initial employment. A comparison of the O/E ratios in Table 3 with those in Table 4 indicates that changing the reference population from US white males to white males of a single industrial population slightly changes most risk ratios, the point estimates of which remain above 1.

[Note: I have computed approximate 95% confidence limits for the ratios, taking *both* observed and expected values as Poisson variables (Hills, 1974), because of the limited size of the population on which the expected figures are based. This obviously produces confidence limits large than if the US population were to be used as reference.]

Unfortunately, no separate figures are provided for different intervals from onset of employment.

In Mancuso's earlier investigation (1970), a subgroup of individuals had been identified at the Ohio plant with clinical case histories of acute bronchitis and pneumonitis due to occupational exposure, as reported by the company to the Health and Safety Laboratory of the New York Operations Office of the Atomic Energy Commission. These reports refer to workers employed and affected in 1940–1948, and form a subgroup of 145 white males among whom 35 deaths had occurred up to 31 December 1967. Six out of the eight (75%) cases of lung cancer deaths that occurred up to that date among the entire Ohio cohort were concentrated in this subgroup, which represents less than 15% of the cohort. Exposure times before the acute respiratory illness were usually short (from a few days to six months).

The working population at the Pennsylvania plant was also investigated by Wagoner *et al.* (1980). The cohort, identified through company records and a medical survey, included all workers at work sometime between 1 January 1942 and 30 August 1968. Table 5 provides some essential information on the composition of the cohort.

Table 5. Vital status as of 31 December 1975 among white males employed in a beryllium production facility

| Status | | |
|---|---|---|
| Known to be alive | 2 101 | |
| Known to be deceased | 875 | |
|    Known to be deceased in USA | | 875 |
|    Death certificates obtained | | 863 |
|    Death certificates outstanding | | 12 |
| Not known to be alive or deceased | 79 | |
| Total | 3 055 | |

A check on a sample of SSA records against the names in the cohort showed that some 6% did not appear in the cohort (they were not, however, added to it).

A few missing ages were replaced with age 20 at entry to work, and missing dates of death were replaced by date of last follow-up (1 January 1976). Vital status ascertainment took place through a variety of channels, including SSA, Internal Revenue Service, etc., with personal contact as a last resort. Causes of death were coded by a qualified nosologist using the ICD revision in effect at the time of death and then converted to ICD, 7th revision. Of the 3055 workers in the cohort, only 17 had worked for more than five years, while some 59% had worked for less than one year (20% less than one month). Mortality from all causes was 875 observed *versus* 816.86 expected (ratio, 1.1; 95% confidence interval, 1.0–1.2). For periods of observation of less than 5, 5–9 and 10 or more years from onset of employment, the ratios were $71/80.49 = 0.9$ (0.7–1.1); $86/102.35 = 0.8$ (0.7–1.0) and $718/634.02 = 1.1$ (1.0–1.2). For all neoplasms, a ratio was found of $143/136.18 = 1.0$ (0.9–1.2), for lung cancer $47/34.29 = 1.4$ (1.0–1.8) [this becomes $47/37.72 = 1.2$ (0.9–1.7) if the expected value is corrected by increasing it by 10%; see Note on p. 206], for non-neoplastic respiratory diseases (excluding influenza and pneumonia) $31/18.76 = 1.7$ (1.1–2.4) and for heart disease $396/349.32 = 1.1$ (1.0–1.2).

Tables 6, 7 and 8 show the breakdown of these figures by time from onset of employment and duration of employment. For lung cancer (ICD 162, 163) there appears to be a rise in ratios with time from first employment, but not with duration of employment: a formally (statistically) significant increase in risk ratio is localized among workers with less than five years of employment observed 25 years after first employment. For non-neoplastic respiratory disease a significantly elevated risk ratio is found in the same group of workers, but the ratios are mostly—as for heart disease —above 1, independently of time since first employment or duration of exposure.

*Beryllium Case Registry*

This Registry was established in 1952 to collect, within the USA, cases of Be-related diseases, and is located in Boston (Hardy *et al.*, 1967; Hasan & Kazemi, 1974). Up to 31 December 1977, 887 cases had been included, of which 212 were acute berylliosis, 44 were chronic manifestations following an acute episode, and 631

Table 6. Observed (O) and expected (E) deaths due to lung cancer according to duration of employment and time since onset of employment among white males employed at some time between January 1942 and September 1968 in a beryllium production facility and followed through 1975

| Interval since onset of employment (years) | Duration of employment (years)[a] | | | | | | | | |
|---|---|---|---|---|---|---|---|---|---|
| | <5 | | | ≥5 | | | Total | | |
| | O/E | Ratio | CI | O/E | Ratio | CI | O/E | Ratio | CI |
| <15 | 8/ 8.74 | 0.9 | 0.4–1.8 | 1/1.63 | 0.6 | 0.1–3.8 | 9/10.37 | 0.9 | 0.4–1.6 |
| 15–24 | 15/12.72 | 1.2 | 0.7–1.9 | 3/2.76 | 1.1 | 0.2–3.5 | 18/15.48 | 1.2 | 0.7–1.8 |
| ≥25 | 17/ 9.98 | 1.7 | 1.0–2.7 | 3/1.89 | 1.6 | 0.4–5.1 | 20/11.87 | 1.7 | 1.0–2.6 |
| Total | 40/31.44 | 1.3 | 0.9–1.7 | 7/6.28 | 1.1 | 0.4–2.3 | 47/37.72 | 1.2 | 0.9–1.7 |

[a] Employment histories ascertained only through 1967–1968
CI, 95% confidence interval

Table 7. Observed (O) and expected (E) deaths due to non-neoplastic respiratory disease (apart from influenza and pneumonia) according to duration of employment and time interval since onset of employment among white males employed at some time between January 1942 and September 1968 in a beryllium production facility and followed through 1975

| Interval since onset of employment (years) | Duration of employment (years)[a] | | | | | | Total | | |
|---|---|---|---|---|---|---|---|---|---|
| | <5 | | | ≥5 | | | | | |
| | O/E | Ratio | CI | O/E | Ratio | CI | O/E | Ratio | CI |
| <15 | 6/ 3.57 | 1.7 | 0.6–3.7 | 1/0.66 | 1.5 | 0.1–8.4 | 7/ 4.23 | 1.7 | 0.7–3.4 |
| 15–24 | 11/ 6.36 | 1.7 | 0.9–3.1 | 1/1.41 | 0.7 | 0.1–4.0 | 12/ 7.77 | 1.6 | 0.8–2.7 |
| ≥25 | 12/ 5.62 | 2.1 | 1.1–3.7 | 0/1.15 | 0.0 | 0 –3.2 | 12/ 6.77 | 1.8 | 0.9–3.1 |
| Total | 29/15.55 | 1.9 | 1.2–2.7 | 2/3.22 | 0.6 | 0.1–2.2 | 31/18.77 | 1.6 | 1.1–2.3 |

[a]Employment histories ascertained only through 1967–1968
CI, 95% confidence interval

Table 8. Observed (O) and expected (E) deaths due to heart disease according to duration of employment and time interval since onset of employment among white males employed at some time between January 1942 and September 1968 in a beryllium production facility and followed through 1975

| Interval since onset of employment (years) | Duration of employment (years)[a] | | | | | | | Total | | |
|---|---|---|---|---|---|---|---|---|---|---|
| | <5 | | | ≥5 | | | | | | |
| | O/E | Ratio | CI | O/E | Ratio | CI | | O/E | Ratio | CI |
| <15 | 109/100.00 | 1.1 | 0.9–1.3 | 23/18.01 | 1.3 | 0.8–1.9 | | 132/118.01 | 1.1 | 0.9–1.3 |
| 15–24 | 124/107.73 | 1.2 | 0.9–1.3 | 37/25.59 | 1.4 | 1.0–2.0 | | 161/133.32 | 1.2 | 1.0–1.4 |
| ≥25 | 85/81.19 | 1.0 | 0.8–1.3 | 18/16.80 | 1.1 | 0.6–1.7 | | 103/97.99 | 1.0 | 0.8–1.2 |
| Total | 318/288.92 | 1.1 | 1.0–1.2 | 78/60.40 | 1.3 | 1.0–1.6 | | 396/349.32 | 1.1 | 1.0–1.2 |

[a] Employment histories ascertained only through 1967–1968
CI, 95% confidence interval

Table 9. Observed (O) and expected (E) deaths from respiratory diseases among white males enrolled in the Beryllium Case Registry while alive, according to beryllium-related respiratory illness indicated at time of entry, at some time between 1 July 1952 and 31 December 1975

| Interval since initial beryllium exposure (years) | Lung cancer | | | Cause of death, non-neoplastic respiratory disease[a] | | |
|---|---|---|---|---|---|---|
| | O/E | Ratio | CI | O/E | Ratio | CI |
| *Acute respiratory illness group (N = 223)* | | | | | | |
| < 15 | 1/0.38 | 2.6 | 0.1–14.7 | 1/0.14 | 7.1 | 0.2– 39.8 |
| ≥ 15 | 5/1.72 | 2.9 | 0.9– 6.8 | 9/0.83 | 10.8 | 5.0– 20.6 |
| Total | 6/2.10 | 2.9 | 1.0– 6.2 | 10/0.97 | 10.3 | 4.9– 18.9 |
| *Chronic respiratory disease group (N = 198)* | | | | | | |
| < 15 | 0/0.15 | 0 | 0 –24.6 | 9/0.05 | 180.0 | 82.3–341.7 |
| ≥ 15 | 1/1.36 | 0.7 | 0.1– 4.1 | 33/0.60 | 55.0 | 37.9– 77.2 |
| Total | 1/1.52 | 0.7 | 0.1– 3.7 | 42/0.65 | 64.6 | 46.6– 87.3 |

[a] Excludes influenza and pneumonia
CI, 95% confidence interval

chronic *ab initio*. Fifty-five cases had been entered between 1972 and 1977 (half of them having been last exposed before 1949). While earlier cases were often among fluorescent lamp workers, later cases came mostly from producers and users of Be alloys and Be oxide ceramics in the nuclear and electronics industries (Sprince & Kazemi, 1980).

Criteria for inclusion in the Registry (which covers only those cases reported to it) are significant exposure to Be, based on a sound history or on the presence of Be in tissues, and objective (physical, X-ray, functional) evidence of lower respiratory tract disease.

Infante *et al.* (1980) followed up the group of 421 white males entered in the Registry while alive between 1 July 1952 and 31 December 1975. Information was extracted from registry records and follow-up procedures; coding for causes of death and data analysis were performed as in the study by Wagoner *et al.* (1980) reviewed above. Vital status could not be ascertained for 64 (15.2%) subjects (for whom person-years were accumulated up to the date they were last seen alive), and for 15 no cause of death could be obtained. Two causes, non-neoplastic respiratory disease (excluding influenza and pneumonia) and lung cancer (ICD 162, 163), show significantly elevated risk ratios. Results for these two causes are shown in some detail in Table 9. Markedly elevated ratios for nonmalignant respiratory causes are found among workers who had experienced acute berylliosis, and even higher ratios for those with chronic berylliosis. Lung cancer shows elevated (by a factor of about three) ratios only in the former group.

## DISCUSSION OF EPIDEMIOLOGICAL STUDIES

Three considerations emerge at the outset. First, the available evidence from several studies concerns, in fact, only two US working populations and a registry of clinical cases covering, in part, these same populations. Second, what the data suggest *prima facie* is not lack of carcinogenicity of Be (as, for example, might be the case if risk ratios fluctuating closely around 1 had been obtained) but an association between lung cancer occurrence and Be-related occupations: the question then pending is whether chance, bias or confounding rather than exposure to Be can explain this association. Third, the association appears to be limited, as far as cancers are concerned, to lung cancer.

The studies of the two cohorts of workers in Ohio and Pennsylvania indicate, in essence, a statistically significant excess risk of lung cancer after long intervals since onset of employment, notably 25 or more years, in workers with less than five (and also less than one) years of duration of employment. The data base of one of the studies, as assembled by Wagoner *et al.* (1980), has been scrutinized closely and criticized, particularly as to the inclusion of a worker who subsequently died of lung cancer and who apparently never actually worked at the plant (Shapley, 1977). However, the fact that two different approaches to cohort identification and follow-up (by Mancuso in 1979 and 1980, and by Wagoner *et al.* in 1980) end with substantially concordant results militates against the play of major biases in subject selection or response assessment. (It should be noted that the final phase of the two studies, i.e., statistical analysis, is not independent, having been performed in collaboration by the authors.) Another fact that goes against biased selection is that elevated risk ratios were also found when an industrial reference population was used instead of the US general population (most ratios were not statistically significant, but this may be due to the relatively small size of the reference population).

After chance and bias, confounding—in particular from smoking—must be considered. A medical survey of workers present in 1968 at the Pennsylvania plant indicated the following distribution of smoking habits, as reported by Wagoner *et al.* (1980): non-smokers, 27.2%; former smokers, 22.4%; smokers of less than one pack/day, 29.0%; smokers of one or more packs/day, 21.4%. This was compared with the distribution of smoking habits in US white males, age-adjusted, at the 1964–1965 health interview survey: non-smokers, 24.7%; former smokers, 20.5%; smokers of less than one pack/day, 39.4%; smokers of one or more packs/day, 15.3%. It can be calculated that the effect of this difference in distribution is to increase (assuming a relative risk of 5 for former smokers, 10 for smokers of less than one pack/day, and 20 for smokers of one or more packs/day) the risk of lung cancer among workers by 4% over the risk in the general population. Such a comparison of *current* smoking habits may not be an effective way of controlling for smoking, but at least it does indicate that unless a substantially grosser imbalance in smoking habits occurred in the past between the workers' population and the general population, the observed increase in lung cancer risk cannot be accounted for by smoking alone.

If, then, the increased risk is, *in toto* or in part, due to exposure, one would expect that the greater the exposure, the higher the risk. Subjects who entered the Beryllium Case Registry, in particular during the 1940s or 1950s, with acute berylliosis had very often experienced high exposures; and, indeed, the Registry data point to an approximately three-fold increase in lung cancer risk in these subjects (Table 9). Also, in the Ohio cohort (Mancuso, 1970), 75% (6/8) of the lung cancer cases observed through 1967 were concentrated in the subgroup of workers (<15% of total) who had experienced acute berylliosis. It cannot be ruled out that some of this excess risk may derive from other exposures associated with berylliosis, for example, diagnostic X-rays, the usage of which ought, however, to have been relatively limited during the short-lasting acute berylliosis.

The position appears to be different, and less clear, for men with chronic berylliosis (Table 9), in whom no excess of lung cancer is detectable. This may be related to a competitive effect, as a cause of death, of chronic berylliosis in respect to lung cancer. Also, chronic berylliosis appears to be related to hypersensitivity, and no dose-response has been established for it, so that it may not be a general marker for high exposures. To complicate matters further, it may incidentally be noted that in sarcoidosis, a condition with features similar to those of chronic berylliosis, one report hints weakly (Brincker & Wilbek, 1974) at a possible increase in lung cancer occurrence. It is also possible that a similar reason, i.e., long employment not implying high exposure, may explain the perplexing finding that the increased lung cancer risk appears to be localized among workers with less than five, and less than one, years of employment. As exposures in the past, particularly up to 1949, were substantially higher, length of employment may be a distorted indicator of the actual exposure accumulated by a worker. This explanation appears to be more plausible than the alternative one, i.e., that workers with relatively short employment are a selected group who experience higher mortality than average (from lung cancer, respiratory diseases, heart diseases) unrelated to Be exposure. That this is unlikely to be the sole explanation is indicated by the persistence of risks above 1 when short-term Be workers are contrasted, not with the general population, but with short-term workers in another industry (Table 4). Also, the excess risks for respiratory diseases and heart diseases can be plausibly interpreted as being related to the action of Be on lungs, and to their cardiac complications (cor pulmonale).

## CONCLUSIONS

In summary, with regard to the qualitative question 'Is beryllium carcinogenic in humans?', my answer is that the most likely single explanation (I stress 'most likely' and 'single') of the ensemble of published epidemiological data is that Be is carcinogenic. The data are inadequate to define a dose (exposure)-response relationship, or possible differential effects of different compounds.

For the sake of completeness, it may be appropriate to append two recent evaluations of Be carcinogenicity made by two international working groups. The first (IARC, 1980) states: 'The epidemiological evidence that occupational exposure to beryllium may lead to an increased lung cancer risk is limited. (One member of the Working Group disassociated himself from this conclusion on the grounds that he considered the epidemiological data sufficient to conclude that beryllium is a confirmed carcinogen in humans.)' The second (Workgroup on Epidemiology, 1981) states: 'The studies demonstrate an excess lung cancer risk following short-term but heavy exposure to beryllium. This same phenomenon was observed for individuals who died from beryllium disease, a clinical endpoint previously demonstrated to be associated with occupational/environmental exposure to beryllium. Our interpretation of the total evidence is that beryllium is the cause of the excess cancer mortality in these groups of employees .... The conclusion that beryllium ... contributes to the development of some cancers in man was reached because this seemed to be the most reasonable interpretation of the available facts. The number of observations in man are small, however ... and it is important to check that the conclusion is correct by continued observation of the cohorts of workers exposed to beryllium in the US.' [Apparently, a study at one Pennsylvania plant is in progress (National Library of Medicine, 1981).]

## REFERENCES

Brincker, H. & Wilbek, E. (1974) The incidence of malignant tumours in patients with respiratory sarcoidosis. *Br. J. Cancer,* **29,** 247–251

Browning, E. (1969) *Toxicity of Industrial Metals,* 2nd ed., London, Butterworths, pp. 67–86

Cotes, J.E., Gilson, J.C., McKerrow, C.B. & Oldham, P.D. (1983) A long-term follow-up of workers exposed to beryllium. *Br. J. ind. Med.,* **40,** 13–21

Documenta Geigy (1972) *Tables Scientifiques,* Basel, Ciba-Geigy, pp. 107–108

Hammond, P.B. & Beliles, R.P. (1980) *Metals.* In: Doull, J., Klaassen, C.D. & Amdur, M.E., eds, *Casarett's and Doull's Toxicology,* New York, Macmillan, pp. 438–439

Hardy, H.L., Rabe, E.W. & Lorch, S. (1967) United States Beryllium Case Registry (1952–1966). Review of its methods and utility. *J. occup. Med.,* **9,** 271–276

Hasan, F.M. & Kazemi, H. (1974) Chronic beryllium disease. A continuing epidemiologic hazard. *Chest,* **65,** 289–293

Hills, M. (1974) *Statistics for Comparative Studies,* London, Chapman & Hall, p. 150

Hueper, W.C. (1966) *Occupational and Environmental Cancers of the Respiratory System,* Berlin, Springer-Verlag, pp. 99–103

IARC (1980) *IARC Monographs on the Evaluation of the Carcinogenic Risk of Chemicals to Humans,* Vol. 23, *Some Metals and Metallic Compounds,* Lyon, pp. 143–204

Infante, P.F., Wagoner, J.K. & Sprince, N.L. (1980) Mortality patterns from lung cancer and nonneoplastic respiratory disease among white males in the Beryllium Case Registry. *Environ. Res.,* **21,** 15–34

MacMahon, B. (1979) *Discussion of Session VII.* In: Lemen, R. & Dement, J.M., eds, *Dusts and Disease,* Park Forest, IL, Pathotox Publishers, pp. 485–487

Mancuso, T.F. (1970) Relation of duration of employment and prior respiratory illness to respiratory cancer among beryllium workers. *Environ. Res.,* **3,** 251–275

Mancuso, T.F. (1979) *Occupational lung cancer among beryllium workers.* In: Lemen, R. & Dement, J.M., eds, *Dusts and Disease,* Park Forest, IL, Pathotox Publishers, pp. 463–472

Mancuso, T.F. (1980) Mortality study of beryllium industry workers' occupational lung cancer. *Environ. Res., 21,* 48–55

Mancuso, T.F. & El-Attar, A.A. (1969) Epidemiological study of the beryllium industry—Cohort methodology and mortality studies. *J. occup. Med., 11,* 422–434

Morgan, W.M.K.C. & Seaton, A. (1975) *Occupational Lung Diseases,* Philadelphia, W.B. Saunders, pp. 223–231

National Library of Medicine (1981) *Tox-Tips,* Bethesda, p. 66–11

Shapley, D. (1977) Occupational cancer: Government challenged in beryllium proceeding. *Science, 198,* 898–901

Smith, R.J. (1981) Beryllium report disputed by listed author. *Science, 211,* 556–557

Sprince, N.L. & Kazemi, H. (1980) US Beryllium Case Registry through 1977. *Environ. Res., 21,* 44–47

Sprince, N.L., Kazemi, H. & Hardy, H.L. (1975) Current (1975) problems of differentiating between beryllium disease and sarcoidosis. *Ann. N.Y. Acad. Sci., 278,* 654–664

Stokinger, H.E. (1981) *Beryllium.* In: Clayton, G.D. & Clayton, F.E., eds, *Patty's Hygiene and Toxicology,* 3rd ed., Vol. 2A, New York, J. Wiley, pp. 1537–1558

Wagoner, J.K., Infante, P.F. & Bayliss, D.L. (1980) Beryllium: an etiologic agent in the induction of lung cancer, nonneoplastic respiratory disease, and heart disease among industrially exposed workers. *Environ. Res., 21,* 15–34

Workgroup on Epidemiology (1981) Problems of epidemiological evidence. Proceedings of a Workshop/Conference on the Role of Metals in Carcinogenesis. *Environ. Health Perspect., 40,* 11–20

# BERYLLIUM:

# CONCLUSION

In the course of discussion, it was reported that the National Institute of Occupational Safety and Hygiene in the USA was examining the original data upon which much of the evidence concerning beryllium and cancer rested, and was adding the experience of some further years of follow-up to it. It was agreed, therefore, to suspend judgement until the results of the re-examination were known. It was thought that the additional data for the period from the end of 1976 would be of particular value.

# GENERAL CONCLUSIONS

# GENERAL CONCLUSIONS

## Summarized by R. DOLL

In the absence of direct human evidence of carcinogenicity, the conclusion that an agent is liable to cause cancer in man is a matter of judgement, in which indirect evidence from other species and laboratory evidence of the effects *in vitro* are weighed together with human evidence of the fate of the agent in the body and observations of the incidence of, and mortality from, cancer in man following exposure to varying amounts in the environment. The conclusion that an agent is not carcinogenic to man, or that, if it is, the risk following exposure to the amounts that are likely to occur is small enough to be disregarded for practical purposes, is arrived at similarly. No hard and fast criteria can be laid down that will automatically lead to an appropriate conclusion in all circumstances. Only one rule is absolute: that all the available evidence must always be taken into account.

Should it ever become possible to make precise quantitative predictions of the risk of cancer in man on the basis of laboratory data, the need for epidemiological observations will diminish. At present, however, such quantitative predictions are extremely unreliable, and management decisions that need to take account of the secondary effects of eliminating or reducing exposure to a specific agent can be greatly assisted by epidemiological studies which can, at least, provide some reasonable estimates of the upper and lower limits of any carcinogenic effect that may be produced. Such evidence provides a sense of proportion which is crucial for determining policies between different lines of action.

The types of epidemiological evidence that can lead to a presumptive negative conclusion and the limitations on their use are discussed in the first two chapters, entitled 'Introduction: Purpose of Symposium', and 'Statistical Considerations'. Examples of the judgements that can be reached in practice, when the totality of the evidence is taken into account, are given in the remaining chapters.

If firmer conclusions are to be reached in the future, it will be necessary for records of exposure to be kept more precisely than they often are now. In industry, records need to be kept not only of the type of work on which each individual is employed, but also of the character and amount of the ambient pollution in different parts of the work place. Epidemiological studies, for their part, need to become more sophisticated, utilizing, whenever possible, biological markers of the extent to which individuals are exposed in practice, to amplify and validate the measures of exposure obtained from histories and records.

Epidemiological studies often need to be carried out on a large scale, particularly if any confidence is to be attached to negative findings, and they are often impracticable without access to the records of hospital admissions, national social security and insurance records, and (in the UK) the National Health Service Central Register. Such access is freely available to *bona fide* researchers in some countries, and it does not appear ever to have been abused. It is to be hoped that similar access will also be made available elsewhere and that the growing and proper concern for the protection of individual privacy will not be allowed to interfere with the release of information to medical research workers in the interest of public health.

The amount of epidemiological data available in any one country is often too small to enable the range of possible effects of any agent to be narrowed sufficiently to be of much practical value and, in this situation, international collaboration is often essential. To be most effective, such collaboration is required in the planning and conduct of epidemiological studies and not only in the final analysis of the results.

# INDEX OF AUTHORS

Acheson, E.D., 91
Anderson, T.W., 119, 165
Armitage, P., 29
Armstrong, B.K., 129, 195
Cabral, J.R.P., 71, 101, 151
Day, N.E., 13
Doll, R., 3, 225
Flamm, W.G., 53, 85, 181, 199
Frankos, V., 85
Fraser, P., 67, 183

Higginson, J., 107, 177
Jensen, O.M., 97
Kinlen, L., 57, 81
Krewski, D., 145
MacMahon, B., 49, 153
Saracci, R., 159, 203
Shubik, P., 33, 125, 163
Vessey, M.P., 37
Wald, N.J., 75

# SUBJECT INDEX

**Aflatoxin,** 191
  and human liver cancer, 114
**Alcohol**
  and human liver cancer, 119
**Allopurinol**
  production from hydrazine, 75
**4-Amino-2-nitrophenol,** 53, 54
**Artificial sweetener**
  (*see also* Saccharin, Cyclamate),
  130–142, 146
**Asbestos,** 16, 17, 21, 25, 92
**Azathioprine,** 7

**Benzene,** 92
**Berylliosis,** 205, 211, 215, 216, 217
**Beryllium,** 199–221
**Bischloro(methyl)ether,** 87–88
**Bladder**
  cancer in humans
    and coffee-drinking, 19, 20
    and formaldehyde, 92, 93, 97
    and hair dyes, 60, 62, 63, 65
    isoniazid, 166, 168–169
    and saccharin/cyclamates, 22, 129–142, 146
  tumours in animals
    and 4-amino-2-nitrophenol, 54
    and cyclamates, 127
    and saccharin, 125–126, 145, 147
**Bone**
  cancer in humans
    and formaldehyde, 98
  osteogenic sarcoma in animals
    and beryllium, 199, 200
**Brain**
  cancer in humans
    and formaldehyde, 92, 93
    and phenobarbital, 153–155, 159–160

**Breast**
  benign disease
    and hair dyes, 62, 65
    and oral contraceptives, 38, 39, 42–43, 47, 49
  cancer in humans
    and DDT, 113–115
    and exogenous oestrogens, 14
    and hair dyes, 58–62, 64, 65
    and isoniazid, 117
    and oral contraceptives, 37–47, 49–50
    and phenobarbital, 154
  tumours in animals (*see* Mammary gland)

**Categories,** of evidence, xi, 9
**Cervix uteri**
  cancer in humans
    and hair dyes, 58, 59, 62, 64, 65, 67
    effect of radiotherapy for, 17–19
  tumours in animals
    and oral contraceptives, 33, 34
**Chlormadinone acetate,** 34, 35, 38
**Colon**
  cancer in humans
    and DDT, 112
    and formaldehyde, 92, 93
    and radiation, 23
**Connective tissue**
  cancer in humans
    and hair dyes, 64
**Corpus uteri**
  cancer in humans
    and hair dyes, 64
**Cyclamates,** 125–148
**Cyclohexylamine,** 127

**DDE,** 101–104, 107, 109, 113
**DDT,** 101–121

# SUBJECT INDEX

Desogestrel, 38
2,4-Diaminoanisole, 53, 55
1,2-Diamino-4-nitrobenzene, 53
1,4-Diamino-2-nitrobenzene, 53
2,4-Diaminotoluene, 53, 54–55
2,5-Diaminotoluene, 53
Dieldrin, 111
Digestive tract
  cancer in humans
    and formaldehyde, 91–92

Endometrium
  cancer in humans
    and DDT, 113, 114
    and hair dyes, 62
Ethinyloestradiol, 33, 34

Formaldehyde, 14, 85–98

Gastrointestinal tract
  cancer in humans
    and hair dyes, 64
Genital organs
  cancer in humans
    and hair dyes, 64
  tumours in animals
    and oral contraceptives, 34

Hair dyes, 53–68
'Healthy worker effect', 15, 29, 81, 92
Hepatitis B virus, 114
β-Hexachlorocyclohexane, 111
Hydrazine, 71–81
Hydrazine sulfate, 163
*IARC Monographs on the Evaluation of the Carcinogenic Risk of Chemicals to Humans*, evaluations, xi, 5, 9, 13, 33, 53, 71, 75, 85, 101, 112, 114, 115, 119, 146, 217

Isoniazid, 71, 75, 163–177
Isonicotinic acid hydrazide (*see* Isoniazid)

Kidney
  cancer in humans
    and formaldehyde, 92, 93, 97
    and isoniazid, 166, 168–169

Leukaemia
  in animals
    and DDT, 113
    and hydrazine, 71
  in humans
    and DDT, 113, 115
    and formaldehyde, 94
    and hair dyes, 59, 62
    and radiation, 17, 19, 21, 22
Liver
  adenomas in animals
    and DDT, 101–102, 113
    and 2-nitro-*para*-phenylenediamine, 54
    and oral contraceptives, 33–35
  adenomas in humans
    and DDT, 115
  cancer in humans
    and DDT, 107, 113–116, 119, 120
    and phenobarbital, 154, 155, 160
  carcinomas in animals
    and DDT, 101–103, 111, 113, 119
    and hydrazine, 71–73
    and hydrazine sulfate, 163
    and isoniazid, 163
    and 2-nitro-*para*-phenylenediamine, 54
    and oral contraceptives, 34
    and phenobarbital, 151, 152, 160
  toxicity
    and isoniazid, 168, 174, 177
Lung
  adenomas in animals
    and isoniazid, 163
  cancer in humans, 13, 14, 24
    and asbestos, 17, 21, 25
    and beryllium, 206–212, 215–217
    and formaldehyde, 93, 97
    and hair dyes, 58, 59, 62
    and hydrazine, 77, 79, 81
    and isoniazid, 167, 168, 170, 173
    and phenobarbital, 156, 157, 159, 160
  carcinomas in animals
    and beryllium, 199–200
    and hydrazine, 71, 72
    and hydrazine sulfate, 163
    and isoniazid, 163
Lymphoma
  in animals
    and DDT, 102, 113
    and hair dyes, 55
    and isoniazid, 163

# SUBJECT INDEX

**Lymphoma (contd)**
    and nitrite, 181
    in humans
        and DDT, 112, 113, 115
        and hair dyes, 62, 64
        and phenobarbital, 156

**Maleic hydrazide,** 71, 72–73
**Mammary gland**
    tumours in animals
        and 2,4-diaminotoluene, 55
        and hydrazine, 72
        and hydrazine sulfate, 163
        and isoniazid, 163
        and oral contraceptives, 33–35
**Medroxyprogesterone acetate,** 34, 35
**Megestrol acetate,** 38
**Mesothelioma**
    in humans
        and asbestos, 16, 17
**Mestranol,** 33, 34, 35
**Mouth**
    cancer in humans
        and formaldehyde, 93
**Mutagenicity**
    of beryllium compounds, 200
    of cyclamate, 127
    of cyclohexylamine, 127
    of DDE, 102
    of DDT, 101, 102
    of formaldehyde, 85, 89
    of hair-dye components, 54, 55
    of hair dyes, 53, 57
    of hydrazine, 71, 72
    of nitrite, 181
    of phenobarbital, 151, 152
    of saccharin, 126, 146

**Nasal cavity**
    cancer in humans
        and formaldehyde, 91–94, 97
        and hydrazine, 77
    tumours in animals
        and formaldehyde, 85, 89, 97
        and hydrazine, 71–72
**Nitrate,** xi, 181–196
**Nitrite,** 181, 182, 183, 195, 196
**Nitrofurazone,** 71
**2-Nitro-*para*-phenylenediamine,** 54

***N*-Nitroso compounds,** 181, 182, 183, 188, 190, 195, 196
**Norethisterone,** 33, 34
**Norethynodrel,** 33, 34
**Norgestrel,** 38

**Oesophagus**
    cancer in humans
        and nitrate, 191, 192
        and nitrite, 191
        and *N*-nitroso compounds, 183, 191
**Oestrogen**
    exogenous, 14, 15
    in oral contraceptives, 33, 38, 40, 49, 115
**Oral contraceptives,** 33–50, 115, 119
**Osteogenic sarcoma** (*see* Bone)
**Ovary**
    cancer in humans
        and hair dyes, 64
        and phenobarbital, 156
    tumours in animals
        and oral contraceptives, 33, 34

**Pancreas**
    cancer in humans
        and phenobarbital, 156
**Pharynx**
    cancer in humans
        and formaldehyde, 93
**Phenobarbital,** 111, 115, 151–160
***meta*-Phenylenediamine,** 53
***para*-Phenylenediamine,** 53
**Phenylhydrazine,** 71
**Phenytoin**
    and phenobarbital, 154
**Pituitary**
    tumours in animals
        and oral contraceptives, 33–35
**Polychlorinated biphenyls,** 111
**Primidone,** 154
**Progestogen**
    in oral contraceptives, 33, 38
**Promotion,** tumour
    by saccharin, 126, 146
    by cyclamate, 127
    by DDT, 111, 112
    by phenobarbital, 151
**Prostate**
    cancer in humans
        and DDT, 113, 114
        and formaldehyde, 92, 93, 97

**Radiotherapy,** 17–23, 167–168
**Respiratory tract**
  cancer in humans
    and formaldehyde, 91, 93, 97
    and hair dyes, 64
    and phenobarbital, 154, 157

**Saccharin,** 15, 22, 125–148
**Skin**
  cancer in humans
    and formaldehyde, 92, 93, 94, 97
    and hair dyes, 64, 67
  tumours in animals
    and 2,4-diaminoanisole, 55
    and hydrazine, 72
**Smoking**
  and beryllium, 216
  and formaldehyde, 93, 98
  and hair dyes, 59, 62, 64
  and hydrazine, 81
  and isoniazid, 167
  and phenobarbital, 156, 157, 159, 160
  and saccharin/cyclamates, 22, 129, 131, 132, 134–141, 147
**Soft-tissue**
  tumours in humans
    and DDT, 112, 114

**Stomach**
  cancer in humans
    and blood group, 23–24
    and hair dyes, 59
    and nitrate, 184–190, 191–192, 195
    and nitrite, 190
    and $N$-nitroso compounds, 183, 190, 192
    and radiation, 18, 20

**TDE,** 101
**Thorium dioxide,** 154
**Thyroid**
  cancer in humans
    and hair dyes, 64
    and phenobarbital, 156
  tumours in animals
    and 2,4-diaminoanisole

**Urinary tract**
  cancer in humans
    and hair dyes, 64

**Vagina**
  cancer in humans
    and vulva, and hair dyes, 62, 65, 67
  tumours in animals
    and oral contraceptives, 33, 34

**X-rays,** 159, 216

# PUBLICATIONS OF THE INTERNATIONAL AGENCY FOR RESEARCH ON CANCER
## SCIENTIFIC PUBLICATIONS SERIES

(Available from Oxford University Press)

No. 1 LIVER CANCER (1971)
176 pages; £ 10.–

No. 2 ONCOGENESIS AND HERPESVIRUSES (1972)
Edited by P.M. Biggs, G. de- Thé & L.N. Payne
515 pages; £ 30.–

No. 3 N-NITROSO COMPOUNDS- ANALYSIS AND FORMATION (1972)
Edited by P. Bogovski, R. Preussmann & E.A. Walker
140 pages; £ 8.50

No. 4 TRANSPLACENTAL CARCINOGENESIS (1973)
Edited by L. Tomatis & U. Mohr
181 pages; £ 11.95

No. 5 PATHOLOGY OF TUMOURS IN LABORATORY ANIMALS. VOLUME 1. TUMOURS OF THE RAT. PART 1 (1973)
Editor-in-Chief V.S. Turusov
214 pages; £ 17.50

No. 6 PATHOLOGY OF TUMOURS IN LABORATORY ANIMALS. VOLUME 1. TUMOURS OF THE RAT. PART 2 (1976)
Editor-in-Chief V.S. Turusov
319 pages; £ 17.50

No. 7 HOST ENVIRONMENT INTERACTIONS IN THE ETIOLOGY OF CANCER IN MAN (1973)
Edited by R. Doll & I. Vodopija
464 pages; £ 30.–

No. 8 BIOLOGICAL EFFECTS OF ASBESTOS (1973)
Edited by P. Bogovski, J.C. Gilson, V. Timbrell & J.C. Wagner
346 pages; £ 25.–

No. 9 N-NITROSO COMPOUNDS IN THE ENVIRONMENT (1974)
Edited by P. Bogovski & E.A. Walker
243 pages; £ 15.–

No. 10 CHEMICAL CARCINOGENESIS ESSAYS (1974)
Edited by R. Montesano & L. Tomatis
230 pages; £ 15.–

No. 11 ONCOGENESIS AND HERPESVIRUSES II (1975)
Edited by G. de-Thé, M.A. Epstein & H. zur Hausen
Part 1, 511 pages; £ 30.–
Part 2, 403 pages; £ 30.–

No. 12 SCREENING TESTS IN CHEMICAL CARCINOGENESIS (1976)
Edited by R. Montesano, H. Bartsch & L. Tomatis
666 pages; £ 30.–

No. 13 ENVIRONMENTAL POLLUTION AND CARCINOGENIC RISKS (1976)
Edited by C. Rosenfeld & W. Davis
454 pages; £ 17.50

No. 14 ENVIRONMENTAL N-NITROSO COMPOUNDS- ANALYSIS AND FORMATION (1976)
Edited by E.A. Walker, P. Bogovski & L. Griciute
512 pages; £ 35.–

No. 15 CANCER INCIDENCE IN FIVE CONTINENTS. VOL. III (1976)
Edited by J. Waterhouse, C.S. Muir, P. Correa & J. Powell
584 pages; £ 35.–

No. 16 AIR POLLUTION AND CANCER IN MAN (1977)
Edited by U. Mohr, D. Schmähl & L. Tomatis
331 pages; £ 30.–

No. 17 DIRECTORY OF ON-GOING RESEARCH IN CANCER EPIDEMIOLOGY 1977 (1977)
Edited by C.S. Muir & G. Wagner
599 pages; out of print

No. 18 ENVIRONMENTAL CARCINOGENS – SELECTED METHODS OF ANALYSIS
Editor-in-Chief H. Egan
Vol. 1 – ANALYSIS OF VOLATILE NITROSAMINES IN FOOD (1978)
Edited by R. Preussmann, M. Castegnaro, E.A. Walker & A.E. Wassermann
212 pages; £ 30.–

No. 19 ENVIRONMENTAL ASPECTS OF N-NITROSO COMPOUNDS (1978)
Edited by E.A. Walker, M. Castegnaro, L. Griciute & R.E. Lyle
566 pages; £ 35.–

No. 20 NASOPHARYNGEAL CARCINOMA: ETIOLOGY AND CONTROL (1978)
Edited by G. de-Thé & Y. Ito
610 pages; £ 35.–

No. 21 CANCER REGISTRATION AND ITS TECHNIQUES (1978)
Edited by R. MacLennan, C.S. Muir, R. Steinitz & A. Winkler
235 pages; £ 11.95

No. 22 ENVIRONMENTAL CARCINOGENS – SELECTED METHODS OF ANALYSIS
Editor-in-Chief H. Egan
Vol. 2 – METHODS FOR THE MEASUREMENT OF VINYL CHLORIDE IN POLY(VINYL CHLORIDE), AIR, WATER AND FOODSTUFFS (1978)
Edited by D.C.M. Squirrell & W. Thain
142 pages; £ 35.–

No. 23 PATHOLOGY OF TUMOURS IN LABORATORY ANIMALS. VOLUME II. TUMOURS OF THE MOUSE (1979)
Editor-in-Chief V.S. Turusov
669 pages; £ 35.–

No. 24 ONCOGENESIS AND HERPESVIRUSES III (1978)
Edited by G. de-Thé, W. Henle & F. Rapp
Part 1, 580 pages; £ 20.–
Part 2, 522 pages; £ 20.–

No. 25 CARCINOGENIC RISKS – STRATEGIES FOR INTERVENTION (1979)
Edited by W. Davis & C. Rosenfeld
283 pages; £ 20.–

No. 26 DIRECTORY OF ON-GOING RESEARCH IN CANCER EPIDEMIOLOGY 1978 (1978)
Edited by C.S. Muir & G. Wagner
550 pages; £ 10.–

No. 27 MOLECULAR AND CELLULAR ASPECTS OF CARCINOGEN SCREENING TESTS (1980)
Edited by R. Montesano, H. Bartsch & L. Tomatis
371 pages; £ 20.–

No. 28 DIRECTORY OF ON-GOING RESEARCH IN CANCER EPIDEMIOLOGY 1979 (1979)
Edited by C.S. Muir & G. Wagner
672 pages; out of print

No. 29 ENVIRONMENTAL CARCINOGENS – SELECTED METHODS OF ANALYSIS
Editor-in-Chief H. Egan
Vol. 3 – ANALYSIS OF POLYCYCLIC AROMATIC HYDROCARBONS IN ENVIRONMENTAL SAMPLES (1979)
Edited by M. Castegnaro, P. Bogovski, H. Kunte & E.A. Walker
240 pages; £ 17.50

No. 30 BIOLOGICAL EFFECTS OF MINERAL FIBRES (1980)
Editor-in-Chief J.C. Wagner
Volume 1, 494 pages; £ 25.–
Volume 2, 513 pages; £ 25.–

No. 31 N-NITROSO COMPOUNDS: ANALYSIS, FORMATION AND OCCURRENCE (1980)
Edited by E.A. Walker, M. Castegnaro, L. Griciute & M. Börzsönyi
841 pages; £ 30.–

No. 32 STATISTICAL METHODS IN CANCER RESEARCH
Vol. 1. THE ANALYSIS OF CASE-CONTROL STUDIES (1980)
By N.E. Breslow & N.E. Day
338 pages; £ 17.50

No. 33 HANDLING CHEMICAL CARCINOGENS IN THE LABORATORY – PROBLEMS OF SAFETY (1979)
Edited by R. Montesano, H. Bartsch, E. Boyland, G. Della Porta, L. Fishbein, R.A. Griesemer, A.B. Swan & L. Tomatis
32 pages; £ 3.95

No. 34 PATHOLOGY OF TUMOURS IN LABORATORY ANIMALS. VOLUME III. TUMOURS OF THE HAMSTER (1982)
Editor-in-Chief V.S. Turusov
461 pages; £ 30.–

No. 35 DIRECTORY OF ON-GOING RESEARCH IN CANCER EPIDEMIOLOGY 1980 (1980)
Edited by C.S. Muir & G. Wagner
660 pages; out of print

No. 36 CANCER MORTALITY BY
OCCUPATION AND SOCIAL CLASS
1851–1971 (1982)
By W.P.D. Logan
253 pages; £ 20.–

No. 37 LABORATORY
DECONTAMINATION AND
DESTRUCTION OF AFLATOXINS $B_1$,
$B_2$, $G_1$, $G_2$ IN LABORATORY WASTES
(1980)
Edited by M. Castegnaro, D.C. Hunt,
E.B. Sansone, P.L. Schuller,
M.G. Siriwardana, G.M. Telling,
H.P. Van Egmond & E.A. Walker
59 pages; £ 5.95

No. 38 DIRECTORY OF ON-GOING
RESEARCH IN CANCER
EPIDEMIOLOGY 1981 (1981)
Edited by C.S. Muir & G. Wagner
696 pages; out of print

No. 39 HOST FACTORS IN HUMAN
CARCINOGENESIS (1982)
Edited by H. Bartsch & B. Armstrong
583 pages; £ 35.–

No. 40 ENVIRONMENTAL
CARCINOGENS. SELECTED
METHODS OF ANALYSIS
Editor-in-Chief H. Egan
Vol. 4. SOME AROMATIC AMINES
AND AZO DYES IN THE GENERAL
AND INDUSTRIAL ENVIRONMENT
(1981)
Edited by L. Fishbein, M. Castegnaro,
I.K. O'Neill & H. Bartsch
347 pages; £ 20.–

No. 41 N-NITROSO COMPOUNDS:
OCCURRENCE AND BIOLOGICAL
EFFECTS (1982)
Edited by H. Bartsch, I.K. O'Neill,
M. Castegnaro & M. Okada
755 pages; £ 35.–

No. 42 CANCER INCIDENCE IN FIVE
CONTINENTS. VOLUME IV (1982)
Edited by J. Waterhouse,
C. Muir, K. Shanmugaratnam & J. Powell
811 pages; £ 35.–

No. 43 LABORATORY
DECONTAMINATION AND
DESTRUCTION OF CARCINOGENS
IN LABORATORY WASTES: SOME
N-NITROSAMINES (1982)
Edited by M. Castegnaro, G. Eisenbrand,
G. Ellen, L. Keefer, D. Klein,
E.B. Sansone, D. Spincer, G. Telling &
K. Webb
73 pages; £ 6.50

No. 44 ENVIRONMENTAL
CARCINOGENS. SELECTED
METHODS OF ANALYSIS
Editor-in-Chief H. Egan
Vol. 5: SOME MYCOTOXINS (1983)
Edited by L. Stoloff, M. Castegnaro,
P. Scott, I.K. O'Neill & H. Bartsch
455 pages; £ 20.–

No. 45 ENVIRONMENTAL
CARCINOGENS. SELECTED
METHODS OF ANALYSIS
Editor-in-Chief H. Egan
Vol. 6: N-NITROSO COMPOUNDS
(1983)
Edited by R. Preussmann, I.K. O'Neill,
G. Eisenbrand, B. Spiegelhalder &
H. Bartsch
508 pages; £ 20.–

No. 46 DIRECTORY OF ON-GOING
RESEARCH IN CANCER
EPIDEMIOLOGY 1982 (1982)
Edited by C.S. Muir & G. Wagner
722 pages; out of print

No. 47 CANCER INCIDENCE IN
SINGAPORE (1982)
Edited by K. Shanmugaratnam, H.P. Lee
& N.E. Day
174 pages; £ 10.–

No. 48 CANCER INCIDENCE IN THE
USSR (1983)
Edited by N.P. Napalkov,
G.F. Tserkovny, V.M. Merabishvili,
D.M. Parkin, M. Smans & C.S. Muir,
75 pages; £ 10.–

No. 49 LABORATORY
DECONTAMINATION AND
DESTRUCTION OF CARCINOGENS
IN LABORATORY WASTES: SOME
POLYCYCLIC AROMATIC
HYDROCARBONS (1983)
Edited by M. Castegnaro, G. Grimmer,
O. Hutzinger, W. Karcher, H. Kunte,
M. Lafontaine, E.B. Sansone, G. Telling &
S.P. Tucker
81 pages; £ 7.95

No. 50 DIRECTORY OF ON-GOING
RESEARCH IN CANCER
EPIDEMIOLOGY 1983 (1983)
Edited by C.S. Muir & G. Wagner
740 pages; out of print

No. 51 MODULATORS IN
EXPERIMENTAL (1983)
Edited by R. Montesano & V.S. Turusov
307 pages; £ 25.–

No. 52 SECOND CANCER IN
RELATION TO RADIATION
TREATMENT FOR CERVICAL
CANCER: RESULTS OF A CANCER
REGISTRY COLLABORATION (1983)
Edited by N.E. Day & J.C. Boice, Jr,
207 pages; £ 17.50

No. 53 NICKEL IN THE HUMAN
ENVIRONMENT (1984)
Edited by F.W. Sunderman, Jr,
529 pages; £ 30.–

No. 54 LABORATORY
DECONTAMINATION AND
DESTRUCTION OF CARCINOGENS
IN LABORATORY WASTES: SOME
HYDRAZINES (1983)
Edited by M. Castegnaro, G. Ellen,
M. Lafontaine, H.C. van der Plas,
E.B. Sansone & S.P. Tucker
87 pages; £ 6.95

No. 55 LABORATORY
DECONTAMINATION AND
DESTRUCTION OF CARCINOGENS
IN LABORATORY WASTES: SOME
N-NITROSAMIDES (1983)
Edited by M. Castegnaro, M. Benard,
L.W. van Broekhoven, D. Fine,
R. Massey, E.B. Sansone, P.L.R. Smith,
B. Spiegelhalder, A. Stacchini, G. Telling
& J.J. Vallon
65 pages; £ 6.95

No. 56 MODELS, MECHANISMS AND
ETIOLOGY OF TUMOUR
PROMOTION (1984)
Edited by M. Börzsönyi, N.E. Day,
K. Lapis & H. Yamasaki
530 pages; £ 30.–

No. 57 N-NITROSO COMPOUNDS:
OCCURRENCE, BIOLOGICAL
EFFECTS AND RELEVANCE TO
HUMAN CANCER (1984)
Edited by I.K. O'Neill, R.C. von Borstel,
C.T. Miller, J. Long & H. Bartsch
1035 pages; £ 75.–

No. 58 AGE-RELATED FACTORS IN
CARCINOGENESIS (1984)
Edited by A. Likhachev, V. Anisimov &
R. Montesano (in press)

No. 59 MONITORING HUMAN
EXPOSURE TO CARCINOGENIC
AND MUTAGENIC AGENTS (1984)
Edited by A. Berlin, M. Draper,
K. Hemminki & H. Vainio (in press)

No. 60 BURKITT'S LYMPHOMA:
A HUMAN CANCER MODEL (1984)
Edited by G. Lenoir, G. O'Conor &
C. Olweny (in press)

No. 61 LABORATORY
DECONTAMINATION AND
DESTRUCTION OF CARCINOGENS
IN LABORATORY WASTES: SOME
HALOETHERS (1984)
Edited by M. Castegnaro, M. Alvarez,
M. Iovu, E.B. Sansone, G.M. Telling &
D.T. Williams
55 pages; £ 5.95

No. 62 DIRECTORY OF ON-GOING
RESEARCH IN CANCER
EPIDEMIOLOGY 1984 (1984)
Edited by C.S. Muir & G. Wagner;
728 pages; £ 18.–

No. 63 VIRUS-ASSOCIATED
CANCERS IN AFRICA (1984)
Edited by A.O. Williams, G.T. O'Conor,
G.B. de-Thé & C.A. Johnson
773 pages; £ 20.–

No. 64 LABORATORY
DECONTAMINATION AND
DESTRUCTION OF CARCINOGENS
IN LABORATORY WASTES: SOME
AROMATIC AMINES AND
4-NITROBIPHENYL (1984)
Edited by M. Castegnaro, J. Barek,
J. Dennis, G. Ellen, M. Klibanov,
M. Lafontaine, R. Mitchum,
P. Van Roosmalen, E.B. Sansone,
L.A. Sternson & M. Vahl (in press)

NON-SERIAL PUBLICATIONS
(Available from IARC)

ALCOOL ET CANCER (1978)
A.J. Tuyns (French only)
42 pages; Fr.fr. 35.–; Sw.fr. 14.–

CANCER MORBIDITY AND CAUSES
OF DEATH AMONG DANISH
BREWERY WORKERS (1980)
O.M. Jensen
145 pages; US$ 25.–; Sw.fr. 45.–

INFORMATION BULLETIN ON THE
SURVEY OF CHEMICALS BEING
TESTED FOR CARCINOGENICITY
No. 8 (1979)
Edited by M.-J. Ghess, H. Bartsch &
L. Tomatis
604 pages; US$ 20.00; Sw.fr. 40.–

No. 9 (1981)
Edited by M.-J. Ghess, J.D. Wilbourn,
H. Bartsch & L. Tomatis
294 pages; US$ 20.00; Sw.fr. 41.–

No. 10 (1982)
Edited by M.-J. Ghess, J.D. Wilbourn &
H. Bartsch
326 pages; US$ 20.00; Sw.fr. 42.–

No. 11 (1984)
Edited by M.-J. Ghess, J.D. Wilbourn,
H. Vainio & H. Bartsch
336 pages; US$ 20.00; Sw.fr. 48.–

## IARC MONOGRAPHS ON THE EVALUATION OF THE
## CARCINOGENIC RISK OF CHEMICALS TO HUMANS

(Available from WHO Sales Agents)

Volume 1, 1972
Some inorganic substances, chlorinated
hydrocarbons, aromatic amines, $N$-nitroso
compounds, and natural products,
184 pages (out of print)

Volume 2, 1973
Some inorganic and organometallic
compounds
181 pages (out of print)

Volume 3, 1973
Certain polycyclic aromatic hydrocarbons
and heterocyclic compounds
271 pages (out of print)

Volume 4, 1974
Some aromatic amines, hydrazine and
related substances, $N$-nitroso compounds
and miscellaneous alkylating agents
286 pages; US$ 7.20; Sw. fr. 18.–

Volume 5, 1974
Some organochlorine pesticides
241 pages (out of print)

Volume 6, 1974
Sex hormones
243 pages; US$ 7.20; Sw.fr. 18.–

Volume 7, 1974
Some anti-thyroid and related substances,
nitrofurans and industrial chemicals
326 pages; US$ 12.80; Sw.fr. 32.–

Volume 8, 1975
Some aromatic azo compounds
357 pages; US$ 14.40; Sw.fr. 36.–

Volume 9, 1975
Some aziridines, $N$, $S$- and $O$-mustards
and selenium
268 pages; US$ 10.80; Sw.fr. 27.–

Volume 10, 1976
Some naturally occurring substances
353 pages; US$ 15.00; Sw.fr. 38.–

Volume 11, 1976
Cadmium, nickel, some epoxides,
miscellaneous industrial chemicals, and
general considerations on volatile
anaesthetics
306 pages; US$ 14.00; Sw.fr. 34.–

Volume 12, 1976
Some carbamates, thiocarbamates and
carbazides
282 pages; US$ 14.00; Sw.fr. 34.–

Volume 13, 1977
Some miscellaneous pharmaceutical
substances
255 pages; US$ 12.00; Sw.fr. 30.–

Volume 14, 1977
Asbestos
106 pages; US$ 6.00; Sw.fr. 14.–

Volume 15, 1977
Some fumigants, the herbicides 2,4-D and
2,4,5-T, chlorinated dibenzodioxins and
miscellaneous industrial chemicals
354 pages; US$ 20.00; Sw.fr. 50.–

Volume 16, 1978
Some aromatic amines and related nitro
compounds – hair dyes, colouring agents
and miscellaneous industrial chemicals
400 pages; US$ 20.00; Sw.fr. 50.–

Volume 17, 1978
Some $N$-nitroso compounds
365 pages; US$ 25.00; Sw.fr. 50.–

Volume 18, 1978
Polychlorinated biphenyls and
polybrominated biphenyls
140 pages; US$ 13.00; Sw.fr. 20.–

Volume 19, 1979
Some monomers, plastics and synthetic
elastomers, and acrolein
513 pages; US$ 35.00; Sw.fr. 60.–

Volume 20, 1979
Some halogenated hydrocarbons
609 pages; US$ 35.00; Sw.fr. 60.–

Supplement No. 1, 1979
Chemicals and industrial processes
associated with cancer in humans (IARC
Monographs 1–20)
71 pages (out of print)

Volume 21, 1979
Sex hormones (II)
583 pages; US$ 35.00; Sw.fr. 60.–

Volume 22, 1980
Some non-nutritive sweetening agents
208 pages; US$ 15.00; Sw.fr. 25.–

Supplement No. 2, 1980
Long-term and short-term screening assays
for carcinogens: a critical appraisal
426 pages; US$ 25.00; Sw.fr. 40.–

Volume 23, 1980
Some metals and metallic compounds
438 pages; US$ 30.00; Sw.fr. 50.–

Volume 24, 1980
Some pharmaceutical drugs
337 pages; US$ 25.00; Sw.fr. 40.–

Volume 25, 1981
Wood, leather and some associated
industries
412 pages; US$ 30.00; Sw.fr. 60.–

Volume 26, 1981
Some antineoplastic and
immunosuppressive agents
411 pages; US$ 30.00; Sw.fr. 62.–

Volume 27, 1982
Some aromatic amines, anthraquinones
and nitroso compounds, and inorganic
fluorides used in drinking-water and
dental preparations
341 pages; US$ 25.00; Sw.fr. 40.–

Volume 28, 1982
The rubber industry
486 pages; US$ 35.00; Sw.fr. 70.–

Volume 29, 1982
Some industrial chemicals and dyestuffs
416 pages; US$ 30.00; Sw.fr. 60.–

Supplement No. 3, 1982
Cross index of synonyms and trade names
in Volumes 1 to 26
199 pages; US$ 30.00; Sw.fr. 60.–

Supplement No. 4, 1982
Chemicals, industrial processes and
industries associated with cancer in
humans (IARC Monographs, Volumes 1
to 29)
292 pages; US$ 30.00; Sw.fr. 60.–

Volume 30, 1983
Miscellaneous pesticides
424 pages; US$ 30.00; Sw.fr. 60.–

Volume 31, 1983
Some food additives, feed additives and
naturally occurring substances
314 pages; US$ 30.00; Sw.fr. 60.–

Volume 32, 1984
Polynuclear aromatic compounds
Part 1, Chemical, environmental and
experimental data
477 pages; US$ 30.00; Sw.fr. 60.–

Volume 33, 1984
Polynuclear aromatic compounds
Part 2, Carbon blacks, mineral oils and
some nitroarene compounds
245 pages; US$ 25.00; Sw.fr. 50.–

Volume 34, 1984
Polynuclear aromatic compounds
Part 3, Some complex industrial exposures
in aluminium production, coal gasification,
coke production, and iron and steel
founding
219 pages; US$ 20.00; Sw.fr. 48.–

THE LIBRARY
UNIVERSITY OF CALIFORNIA
San Francisco
(415) 476-2335

**THIS BOOK IS DUE ON THE LAST DATE STAMPED BELOW**

Books not returned on time are subject to fines according to the Library Lending Code. A renewal may be made on certain materials. For details consult Lending Code.

14 DAY
JUN 30 1986
RETURNED
JUN 25 1986
14 DAY
AUG 19 1986
RETURNED
AUG 21 1986
**14 DAY**
**JAN -2 1987**
RETURNED
DEC 18 1986

14 DAY
OCT 21 1987
RETURNED
OCT 22 1987
14 DAY
DEC 13 1987
RETURNED
DEC 15 1987
14 DAY
FEB 16 1989
RETURNED
FEB -6 1989